TWENTY FIRST CENTURY
SCIENCE

Project Directors

Angela Hall   Emma Palmer

Robin Millar   Mary Whitehouse

Author

Philippa Gardom Hulme

D0453101

THE UNIVERSITY *of* York

THE SALTERS' INSTITUTE

Nuffield Foundation

OCR
RECOGNISING ACHIEVEMENT

OXFORD
UNIVERSITY PRESS

Official Publisher Partnership

OXFORD
UNIVERSITY PRESS

Great Clarendon Street, Oxford OX2 6DP

Oxford University Press is a department of the University of Oxford.
It furthers the University's objective of excellence in research,
scholarship, and education by publishing worldwide in

Oxford   New York

Auckland Cape Town Dar es Salaam Hong Kong Karachi
Kuala Lumpur Madrid Melbourne Mexico City Nairobi
New Delhi Shanghai Taipei Toronto

With offices in

Argentina Austria Brazil Chile Czech Republic France Greece
Guatemala Hungary Italy Japan Poland Portugal Singapore
South Korea Switzerland Thailand Turkey Ukraine Vietnam

Oxford is a registered trade mark of Oxford University Press
in the UK and in certain other countries.

British Library Cataloguing in Publication Data.

Data available.

ISBN 978-0-19-913840-1

10 9 8 7 6 5 4 3 2 1

Printed in Great Britain by Bell and Bain Ltd, Glasgow.

Paper used in the production of this book is a natural, recyclable product made
from wood grown in sustainable forests. The manufacturing process conforms to
the environmental regulations of the country of origin.

Acknowledgements
Illustrations by IFA Design, Plymouth, UK, Clive Goodyer, and Q2A Media.

Author acknowledgements
Many thanks to Catherine and Sarah for checking the puzzles, and to Barney for
his inspirational ideas. Thanks to Ruth for her careful editing, and to Les, Sophie,
and Barry at OUP for all their help and patience.

Introduction

About this book

Welcome to the Twenty-First Century Chemistry Revision Guide! This book will help you prepare for all your GCSE Chemistry module tests. There is one section for each of the chemistry modules C1–C7, as well as six sections covering Ideas about science. Each section includes several types of pages to help you revise.

Workout: These are to help you find out what you can remember already, and get you thinking about the topic. They include puzzles, flow charts, and lots of other types of questions. Work through these on your own or with a friend, and write your answers in the book. If you get stuck, look in the Factbank. The index will help you find what you need. Check your answers in the back of the book.

Factbank: The Factbanks summarise information from the module in just a few pages. For C1–C7, the Factbanks are divided into short sections, each linked to different statements in the Specification. The Ideas about science Factbanks are different. They are conversations, covering the ideas you will need to apply in different contexts. Read them aloud with a friend if you want to.

Quickfire: Sections C1–C7 each have Quickfire questions. These are short questions that cover most of the content of the module. For some questions, there is space to answer in the book. For others, you will need to use paper or an exercise book.

GCSE-style questions: These are like the questions in the module tests. You could work through them using the Factbank to check things as you go, or do them under test conditions. The answers are in the back of the book. Most sections include one 6-mark question, designed to test your ability to organise ideas, and write in clear and correct English. Use these to help you practise for this type of question in the module tests.

In every section, content covered at Higher-tier only is shown like this.

Other help: This page and the next one include vital revision tips and hints to help you work out what questions are telling you to do. Don't skip these!

Making the most of revision

Remember, remember: You probably won't remember much if you just read this book. Here are some suggestions to help you revise effectively.

Plan your time: Work out how many days there are before your test. Then make a timetable so you know which topics to revise when. Include some time off.

Revise actively, don't just read the Factbanks. Highlight key points, scribble extra details in the margin or on Post-it notes, and make up ways to help you remember things. The messier the Factbanks are by the time you take your tests, the better!

Mind maps: try making mind maps to summarise the information in a Factbank. Start with an important idea in the middle. Use arrows to link this to key facts, examples, and other science ideas.

Test yourself on key facts and ideas. Use the Quickfire sections in this book, or get a friend to ask you questions. You could make revision cards, too. Write a question on one side, and the answer on the other. Then test yourself.

Try making up songs or rhymes to help you remember things. You could make up **mnemonics**, too, like this one for the gases in the Earth's atmosphere:

Never **O**ffend **A** **C**ockroach.

Apply your knowledge: Don't forget you will need to apply knowledge to different contexts, and evaluate data and opinions. The GCSE-style questions in this book give lots of opportunities to practise these skills. Your teacher may give you past test papers, too.

Ideas about science: should not be ignored. These are vital. In your module tests, there could be questions on any of the Ideas about science you have covered so far, set in the context of most of the topics you have covered.

Take short breaks: take plenty of breaks during revision – about 10 minutes an hour works for most people. It's best not to sit still and relax in your breaks – go for a walk, or do some sport. You'll be surprised at what you can remember when you come back, and at how much fresher your brain feels!

Answering exam questions

Read the question carefully, and find the command word. Then look carefully at the information in the question, and at any data. How will they help you answer the question? Use the number of answer lines and the number of marks to help you work out how much detail the examiner wants.

Then write your answer. Make it easy for the examiner to read and understand. If a number needs units, don't forget to include them.

Six-mark questions

Follow the steps below to gain the full six marks:
- Work out exactly what the question is asking.
- Jot down key words to help your answer.
- Organise the key words. You might need to group them into advantages and disadvantages, or sequence them to describe a series of steps.
- Write your answer. Use the organised key words to help.
- Check and correct your spelling, punctuation, and grammar.

Below are examiner's comments on two answers to the question: ***"Outline the arguments for and against recycling metals such as aluminium, compared to extracting them from their ores."***

✎ The quality of written communication will be assessed.

Command words

Calculate Work out a number. Use your calculator if you like. You may need to use an equation.
Compare Write about the ways in which two things are the same, and how they are different.
Describe Write a detailed answer that covers what happens, when it happens, and where it happens. Your answer must include facts, or characteristics.
Discuss Write about the issues, giving arguments for and against something, or showing the difference between ideas, opinions, and facts.
Estimate Suggest a rough value, without doing a complete calculation. Use your science knowledge to suggest a sensible answer.
Explain Write a detailed answer that says how and why things happen. Give mechanisms and reasons.
Evaluate You will be given some facts, data, or an article. Write about these, and give your own conclusion or opinion on them.
Justify Give some evidence or an explanation to tell the examiner why you gave an answer.
Outline Give only the key facts, or the steps of a process in the correct order.
Predict Look at the data and suggest a sensible value or outcome. Use trends in the data and your science knowledge to help you.
Show Write down the details, steps, or calculations to show how to get an answer.
Suggest Apply something you have learnt to a new context, or to come up with a reasonable answer.
Write down Give a short answer. There is no need for an argument to support your answer.

Answer	Examiners' comments
Alluminnium is expensive and uses lots of emergy too get it. So it is better to resicle it but you cood get it from the ground and youse electrisity but it needs lots of electrisity. And my dad says he cant be bovvered to recycle his cans.	**Grade G** answer: this makes some correct points. However, the points are not well organised and it is not clear which arguments are for and which against recycling metals. There are mistakes of spelling, grammar, and punctuation.
Extracting aluminium from its ore requires much electrical energy. The process produces carbon dioxide (a greenhouse gas) and red mud waste, which damages the environment. There is only a limited amount of aluminium ore in the world, so once it is used up there will be none for people in future. *On the other hand, recycling aluminium requires less energy. If we recycle, there will be more aluminium ore left for future generations, so it is more sustainable to recycle. Some people think recycling is a nuisance, but in my opinion it is worth the extra effort.*	**Grade A/A*** answer: the arguments are made clearly and are organised logically. The candidate has referred to the idea of sustainability. The spelling, punctuation, and grammar are faultless.

Equations, units, and data
Equations

C6 Chemical synthesis

$$\text{Percentage yield} = \frac{\text{actual yield}}{\text{theoretical yield}} \times 100\%$$

C7 Further chemistry

$$\text{Concentration of a solution} = \frac{\text{mass of solute}}{\text{volume of solution}}$$

$$\text{Retardation factor } (R_f) = \frac{\text{distance travelled by solute}}{\text{distance travelled by solvent}}$$

Units

Length: metres (m), kilometres (km), centimetres (cm), millimetres (mm), micrometres (μm), nanometres (nm)

Mass: kilograms (kg), grams (g), milligrams (mg)

Time: seconds (s), milliseconds (ms)

Temperature: degrees Celsius (°C)

Area: cm^2, m^2

Volume: cm^3, dm^3, m^3, litres (l), millilitres (ml)

Data

You will be given these data in an examination. You do not need to learn them.

C5 Chemicals of the natural environment

Dry air is made up of approximately 78% nitrogen, 21% oxygen, 1% argon, and 0.04% carbon dioxide.

Chemical formulae

C1

Name	Formula
carbon dioxide	CO_2
carbon monoxide	CO
sulfur dioxide	SO_2
nitrogen monoxide	NO
nitrogen dioxide	NO_2
water	H_2O

C4

water	H_2O
hydrogen	H_2
chlorine	Cl_2
bromine	Br_2
iodine	I_2
lithium chloride	$LiCl$
sodium chloride	$NaCl$
potassium chloride	KCl
lithium bromide	$LiBr$
sodium bromide	$NaBr$
potassium bromide	KBr
lithium iodide	LiI
sodium iodide	NaI
potassium iodide	KI

C5

nitrogen	N_2
oxygen	O_2
argon	Ar
carbon dioxide	CO_2
sodium chloride	$NaCl$
magnesium chloride	$MgCl_2$
sodium sulfate	Na_2SO_4
magnesium sulfate	$MgSO_4$
potassium chloride	KCl
potassium bromide	KBr

C6

chlorine	Cl_2
hydrogen	H_2
nitrogen	N_2
oxygen	O_2
hydrochloric acid	HCl
sulfuric acid	H_2SO_4
nitric acid	HNO_3
sodium hydroxide	NaOH
sodium chloride	NaCl
sodium carbonate	Na_2CO_3
sodium nitrate	$NaNO_3$
sodium sulfate	Na_2SO_4
potassium chloride	KCl
magnesium oxide	MgO
magnesium hydroxide	$Mg(OH)_2$
magnesium carbonate	$MgCO_3$
magnesium chloride	$MgCl_2$
magnesium sulfate	$MgSO_4$
calcium carbonate	$CaCO_3$
calcium chloride	$CaCl_2$
calcium sulfate	$CaSO_4$

C7

methanol	CH_3OH
ethanol	C_2H_5OH
methanoic acid	HCOOH
ethanoic acid	CH_3COOH

Tests for ions

Tests for positive ions

Ion	Test	Observation
calcium Ca^{2+}	add dilute sodium hydroxide solution	white precipitate, insoluble in excess sodium hydroxide solution
copper Cu^{2+}	add dilute sodium hydroxide solution	blue precipitate
iron(II) Fe^{2+}	add dilute sodium hydroxide solution	green precipitate
iron(III) Fe^{3+}	add dilute sodium hydroxide solution	red-brown precipitate
zinc Zn^{2+}	add dilute sodium hydroxide solution	white precipitate, soluble in excess sodium hydroxide solution

Tests for negative ions

Ion	Test	Observation
carbonate CO_3^{2-}	add dilute acid	fizzes, carbon dioxide gas produced
chloride Cl^-	add dilute nitric acid, then silver nitrate solution	white precipitate
bromide Br^-	add dilute nitric acid, then silver nitrate solution	cream precipitate
iodide I^-	add dilute nitric acid, then silver nitrate solution	yellow precipitate
sulfate SO_4^{2-}	add dilute acid, then barium chloride or barium nitrate solution	white precipitate

The Periodic Table of the Elements

1	2							3	4	5	6	7	0
													4 **He** helium 2
7 **Li** lithium 3	9 **Be** beryllium 4							11 **B** boron 5	12 **C** carbon 6	14 **N** nitrogen 7	16 **O** oxygen 8	19 **F** fluorine 9	20 **Ne** neon 10
23 **Na** sodium 11	24 **Mg** magnesium 12							27 **Al** aluminium 13	28 **Si** silicon 14	31 **P** phosphorus 15	32 **S** sulfur 16	35.5 **Cl** chlorine 17	40 **Ar** argon 18
39 **K** potassium 19	40 **Ca** calcium 20	45 **Sc** scandium 21	48 **Ti** titanium 22	51 **V** vanadium 23	52 **Cr** chromium 24	55 **Mn** manganese 25	56 **Fe** iron 26	59 **Co** cobalt 27	59 **Ni** nickel 28	63.5 **Cu** copper 29	65 **Zn** zinc 30	70 **Ga** gallium 31	73 **Ge** germanium 32
86 **Rb** rubidium 37	88 **Sr** strontium 38	89 **Y** yttrium 39	91 **Zr** zirconium 40	93 **Nb** niobium 41	96 **Mo** molybdenum 42	98 **Tc** technetium 43	101 **Ru** ruthenium 44	103 **Rh** rhodium 45	106 **Pd** palladium 46	108 **Ag** silver 47	112 **Cd** cadmium 48	115 **In** indium 49	119 **Sn** tin 50
133 **Cs** caesium 55	137 **Ba** barium 56	139 **La*** lanthanum 57	178 **Hf** hafnium 72	181 **Ta** tantalum 73	184 **W** tungsten 74	186 **Re** rhenium 75	190 **Os** osmium 76	192 **Ir** iridium 77	195 **Pt** platinum 78	197 **Au** gold 79	201 **Hg** mercury 80	204 **Tl** thallium 81	207 **Pb** lead 82
[223] **Fr** francium 87	[226] **Ra** radium 88	[227] **Ac*** actinium 89	[261] **Rf** rutherfordium 104	[262] **Db** dubnium 105	[266] **Sg** seaborgium 106	[264] **Bh** bohrium 107	[277] **Hs** hassium 108	[268] **Mt** meitnerium 109	[271] **Ds** damstadtium 110	[272] **Rg** roentgenium 111			

Additional group 5, 6, 7 entries:

5	6	7
75 **As** arsenic 33	79 **Se** selenium 34	80 **Br** bromine 35
(Kr: 84 **Kr** krypton 36)		
122 **Sb** antimony 51	128 **Te** tellurium 52	127 **I** iodine 53
(131 **Xe** xenon 54)		
209 **Bi** bismuth 83	[209] **Po** polonium 84	[210] **At** astatine 85
(222 **Rn** radon 86)		

Key

relative atomic mass
atomic symbol
name
atomic (proton) number

Example:

1
H
hydrogen
1

Elements with atomic numbers 112–116 have been reported but not fully authenticated

The lanthanoids (atomic numbers 58–71) and the actinoids (atomic numbers 90–103) have been omitted.

1 Label the pie chart with the names of the gases of the Earth's atmosphere.

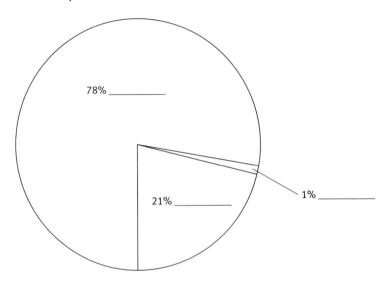

78% _____

21% _____

1% _____

2 Complete the captions to show how the Earth's atmosphere may have been formed.

The early atmosphere was mainly _____ and _____.

Water vapour _____ to form oceans. Carbon dioxide _____ in the oceans. Later it formed _____ rocks.

Early plants removed _____ from the atmosphere by photosynthesis, and added _____ to the atmosphere.

H 3 Write these formulas in sensible places on the drawings.

O_2 NO_2 CO_2 N_2 H_2O NO   C   CO

petrol (hydrocarbons)

My car has a catalytic converter.

A

gases out

gases in

petrol (hydrocarbons)

B

Cats didn't exist when they made this old banger!

gases out

gases in

4 Fill in the gaps.

Octane is a hydrocarbon. It is a compound that contains _____ and _____

only. When it burns, it reacts with _____ from the air to make _____

_____ and _____. This is a _____ reaction. There

are the same total _____ of atoms of each element in the reactants and in the

_____ . The atoms are joined together differently after the reaction; they have

been _____ . The properties of the _____ are different from the

properties of the reactants.

H The total mass of reactants is the same as the total mass of _____ .

5 Fill in the empty boxes to summarise some combustion reactions of coal.

	Reactants		Product
Name	coal (with no sulfur impurities)	oxygen (from a plentiful supply of air)	
Formula	C	O_2	CO_2
Diagram			

	Reactants		Products		
Name	coal (with no sulfur impurities)	oxygen (from a limited supply of air)		carbon monoxide	particulate carbon
Formula	C		CO_2		C
Diagram					

	Reactants		Products	
Name	coal (with sulfur impurities)	oxygen (from a plentiful supply of air)		
Formula	C and S		CO_2	SO_2
Diagram				

H **6** Write in the speech bubbles to finish what the people are saying.

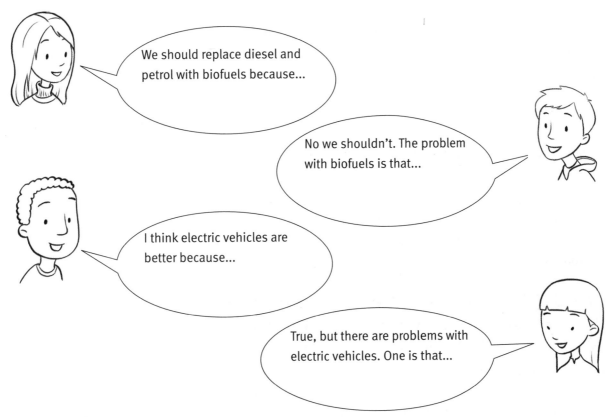

We should replace diesel and petrol with biofuels because...

No we shouldn't. The problem with biofuels is that...

I think electric vehicles are better because...

True, but there are problems with electric vehicles. One is that...

C 1

7 Fill in the empty boxes.

Pollutant name	Pollutant formula	Where the pollutant comes from	Problems the pollutant causes	One way of reducing the amount of this pollutant added to the atmosphere
sulfur dioxide				
nitrogen oxides				
carbon dioxide				
carbon monoxide				
particulate carbon				

C1.1.1–4 What is in the atmosphere?

The **atmosphere**, or air, is the layer of gases surrounding the Earth. It is a mixture of gases, each made up of small **molecules**. A molecule is a group of atoms joined together. In air, there are big spaces between the molecules. Clean air consists of:

- 78% nitrogen
- 21% oxygen
- 1% argon
- small amounts of other gases, like water vapour and carbon dioxide.

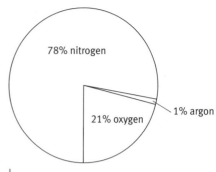

This pie chart shows the percentages of the main gases in clean air.

Human activity and volcanic eruptions add other gases and **particulates** to the air. Particulates are tiny pieces of solid. Many of these gases and particulates are **pollutants**. They affect air quality.

C1.1.5–8 How was the atmosphere formed?

The Earth's early atmosphere was probably mainly carbon dioxide and water vapour. These gases came from volcanoes.

The Earth cooled. Water vapour condensed to make oceans. Some carbon dioxide dissolved in the oceans, later making sedimentary rocks. Some of the carbon from the carbon dioxide ended up in fossil fuels.

Over time, the amount of oxygen in the air increased. This may have been the result of photosynthesis in early plants. The plants took carbon dioxide from the atmosphere, and added oxygen to the atmosphere.

C1.2.1–3, C1.2.7–9 How do reactions make air pollutants?

Coal is mainly **carbon**. Petrol, diesel, and fuel oil are mainly **hydrocarbons**. Hydrocarbons are compounds of hydrogen and carbon only. These **fossil fuels** burn in vehicles like cars and trains, and in power stations to generate electricity.

When fossil fuels burn they react with oxygen from the air. The main products are carbon dioxide and water (hydrogen oxide). Burning reactions are also called **combustion** reactions.

In all chemical reactions, the atoms are rearranged. The same atoms make up both the reactants and the products; they are just joined together differently.

This means that, in a chemical reaction, the total mass of products is equal to the total mass of reactants. Mass is conserved.

In all chemical reactions, the properties of the reactants and products are different.

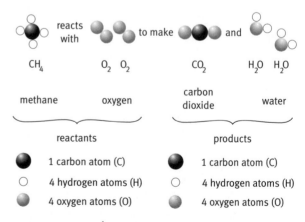

Methane (natural gas) reacts with oxygen to make carbon dioxide and water.

C1.2.4–6 What are oxidation and reduction reactions?

Oxidation reactions add oxygen to a chemical. The burning reactions of fuels are examples of oxidation reactions.

Fuels burn more quickly in pure oxygen than in air. Oxygen can be separated from the air. The oxygen can be used to support combustion reactions in oxy-fuel welding torches.

C1

C1.9.1, C1.2.11–14 Where do air pollutants come from?

Burning fossil fuels make carbon dioxide. Sometimes there is not enough oxygen to convert all the carbon in the fuel to carbon dioxide. Then **incomplete burning** occurs. This makes carbon monoxide gas and particulate carbon.

Some fossil fuels contain sulfur. When the fuel burns, the sulfur reacts with oxygen to make sulfur dioxide.

At high temperatures inside vehicle engines, nitrogen and oxygen from the air react to make nitrogen oxides.

Name of pollutant	Formula	Diagram of molecule
carbon dioxide	CO_2	
carbon monoxide	CO	
sulfur dioxide	SO_2	
nitrogen monoxide	NO	
nitrogen dioxide	NO_2	
water	H_2O	

H Nitrogen monoxide (NO) forms first. This reacts with more oxygen from the air to make nitrogen dioxide (NO_2). Together, NO and NO_2 are called NO_x.

The table shows the formulae of some products of burning reactions.

C1.1.10, C1.2.15 What problems do air pollutants cause?

Atmospheric pollutants do not just disappear. Some carbon dioxide from burning fossil fuels is used for photosynthesis. Some dissolves in rainwater and in the sea. The rest of the carbon dioxide remains in the atmosphere.

Some pollutants harm humans directly, including carbon monoxide. This reduces the amount of oxygen your blood can carry.

Some pollutants harm the environment. They may harm humans indirectly, including:
- Sulfur dioxide and oxides of nitrogen react with water and oxygen to make acid rain.
- Carbon dioxide contributes to climate change.
- Particulate carbon makes surfaces dirty.

> **Exam tip**
>
> Practise using the formulae in the table to draw the molecule diagrams.

C1.3.1–5 How can we improve air quality?

The only way of making less carbon dioxide is to burn smaller amounts of fossil fuels.

Power stations

We can reduce air pollution from fossil fuel power stations by:
- using less electricity
- removing sulfur impurities from natural gas and oil before burning
- removing sulfur dioxide and particulates from flue gases (the gases that power stations emit).

Ⓗ Sulfur dioxide is acidic. It is removed from flue gases by reacting it with an alkali in **wet scrubbing**. Two alkaline substances are used:
- Seawater – the flue gases are sprayed with seawater. Substances in the seawater react with sulfur dioxide.
- A slurry of calcium oxide powder and water – calcium oxide and sulfur dioxide react to make calcium sulfate. The calcium sulfate is used to make building plaster.

Vehicles

We can reduce air pollution from vehicle exhaust gases by:
- developing efficient engines that burn less fuel
- using low sulfur fuels
- using catalytic converters in vehicles to
  - add oxygen to carbon monoxide in an oxidation reaction. The product is carbon dioxide.
  - remove oxygen from nitrogen monoxide in a **reduction** reaction. The products are nitrogen and oxygen.
- using public transport instead of cars
- having legal limits on exhaust emissions, enforced by MOT tests
- fuelling vehicles with biofuels, or using electric vehicles.

Ⓗ **Biofuels** are made from plants like sugar cane. The plants take carbon dioxide from the atmosphere as they grow. When biofuels burn, carbon dioxide returns to the atmosphere. Biofuel plants grow on land that could be used for food crops.

Electric vehicles are powered by batteries. They do not produce waste gases during use. However, the electricity may have been generated by burning fossil fuels. Electric vehicles need recharging often.

Use extra paper to answer these questions if you need to.

1 Draw lines to link each pollutant to a problem it causes.

Pollutants
sulfur dioxide
carbon dioxide
carbon monoxide
particulate carbon

Problems
makes surfaces dirty
acid rain
reduces the amount of oxygen the blood carries
climate change

2 Complete the table to show the percentage of each gas in clean air.

Gas	Percentage in air
argon	
nitrogen	
oxygen	

3 Choose words from the box to fill in the gaps in the sentences below. The words in the box may be used once, more than once, or not at all.

> **carbon   nitrogen   oxygen   carbon dioxide**
>
> **hydrogen   carbon monoxide   hydrocarbons**

Coal contains mainly _____ . Petrol and diesel are mainly _____ . These are compounds made up of _____ and carbon only. When petrol burns, its atoms react with atoms from _____ molecules. The products of the reaction are mainly _____ and water.

4 Highlight the statements below that are **true**. Then write corrected versions of the statements that are **false**.
  a The air is a mixture of gases.
  b The spaces between molecules in the air are small.
  c Carbon monoxide is indirectly harmful to humans.
  d Nitrogen dioxide is indirectly harmful to humans.
  e Fuels burn more slowly in air than in pure oxygen.

5 The statements below describe some changes in the atmosphere. Write the letters of the steps in the order that they probably happened.
  A Burning fossil fuels added extra carbon dioxide to the atmosphere.
  B Water vapour condensed to form the oceans.
  C Volcanoes added carbon dioxide and water vapour to the atmosphere.
  D Early plants added oxygen to the atmosphere.

6 Explain how carbon dioxide gas was removed from the early atmosphere.

7 Highlight the one correct word or phrase in each pair of **bold** words.

Methane is a **hydrocarbon/particulate**. When it burns, it reacts with **nitrogen/oxygen** from the air in an **oxidation/reduction** reaction. In **oxidation/reduction** reactions, oxygen is lost from a substance.

8 Copy and complete the table.

Formula	Diagram of molecule	Name
CO		
		sulfur dioxide
		carbon dioxide
NO_2		
H_2O		
		nitrogen monoxide

9 List five ways of reducing the atmospheric pollution caused by exhaust emissions from motor vehicles.

10 The diagram below shows the molecules in the reaction of a hydrocarbon fuel, methane, with oxygen. Describe what the diagram tells you about the reaction.

H 11 Fill in the gaps in the sentences below.
At the high temperature of car engines, oxygen and _____ from the air react to make _____ _____ gas, with the formula _____ . This then reacts with more oxygen to make _____ _____ gas, with the formula _____ . Together, these oxides of nitrogen are referred to as _____ . Oxides of nitrogen cause _____ _____ .

12 Describe two ways of removing sulfur dioxide from power station flue gases by wet scrubbing.

13 List the benefits and problems of using biofuels to fuel cars.

14 Identify the advantages and disadvantages of electric cars, compared with diesel cars.

15 A scientist found that 3 g of a hydrocarbon used up 8 g of oxygen when it burned completely. Calculate the mass of carbon dioxide produced.

16 A scientist burned 12 g of carbon in pure oxygen. The reaction produced 44 g of carbon dioxide. Calculate the mass of oxygen that reacted with the carbon.

17 1 g of sulfur reacts with 1 g of oxygen. What mass of sulfur dioxide is made?

18 A scientist burned 4 g of methane in 16 g of oxygen. She made 11 g of carbon dioxide, and some water vapour. Calculate the mass of water vapour.

1 Menna does an experiment to estimate the percentage of oxygen in the air.

She heats some copper strongly whilst passing air over it, backwards and forwards, from two syringes.

The oxygen in the air in the syringes reacts with the copper.

Menna works out the decrease in the volume of air.

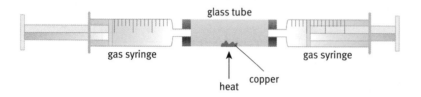

	Run 1	Run 2	Run 3	Run 4
Volume of air in syringes at start (cm³)	100	100	100	100
Volume of air in syringes at end (cm³)	82	86		84
Decrease in volume (cm³)	18		15	16

a Calculate the missing volumes.
   Write your answers in the table. [2]

b Select data from the table to calculate the mean decrease in volume.

   Mean volume = _____ cm³ [2]

c The expected decrease in volume is 21 cm³. Explain why.

   _____ [1]

d Suggest an explanation for the difference between the expected decrease in volume and the decrease in volume you calculated in part (b).

   _____

   _____ [1]

Total [6]

2 a i Sulfur dioxide is made when sulfur reacts with oxygen from the air.
Finish the diagram to represent this reaction.

[1]

ii One source of sulfur is in the coal that power stations burn to generate electricity.
Name one other source of the sulfur that reacts to make sulfur dioxide.

_____ [1]

b The graph shows how the amount of sulfur dioxide emitted by China changed in the 1980s.

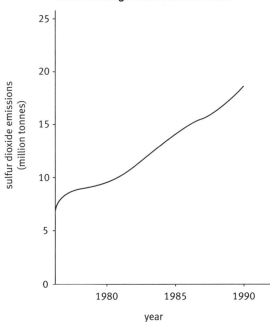

Sulfur dioxide gas emissions from China

i Use the graph to complete the sentence below.

Between 1980 and 1990 the amount of sulfur dioxide emitted by China _____ .[1]

ii Suggest why the amount of sulfur dioxide emitted by China changed in this way.

_____ [1]

c i   Sulfur dioxide mixes with other gases in the atmosphere.
Circle the names of the two gases in the Earth's atmosphere that are present in the two largest amounts.

**nitrogen   carbon dioxide   oxygen   argon
hydrogen**   [2]

ii   In the atmosphere, sulfur dioxide reacts with water to make acid rain.
Why is acid rain a problem?
Put ticks in the three correct boxes.

Acid rain damages buildings made
of limestone.   ☐

Acid rain makes lakes more acidic.   ☐

Acid rain damages trees.   ☐

Acid rain increases the pH of streams.   ☐   [2]

d   This graph shows how the amount of sulfur dioxide emitted by the UK has changed since 1980.

Describe the trend shown by the graph.   [1]

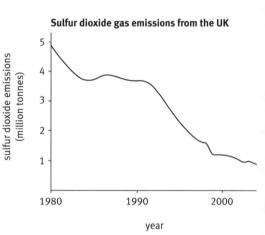

Sulfur dioxide gas emissions from the UK

Total [9]

Going for the highest grades

3 Cars burn hydrocarbon fuels. Carbon dioxide is a product of these burning reactions. The table shows the masses and carbon dioxide emissions of some Volkswagen cars.

Car model	Minimum car mass (kg)	Average CO$_2$ emissions in g / km
Fox	978	144
Polo	1000	138
New Golf	1142	149
Touran	1423	176
Toureg	2214	324

A student says that the data shows that heavier cars cause increased carbon dioxide emissions.

Do you think the student is correct? Use data from the table, as well as your own knowledge and understanding, to support your decision.

✎ The quality of written communication will be assessed in your answer to this question.

Write your answer on separate paper or in your exercise book.

Total [6]

1 Look at the drawing of a child's tricycle.
  Fill in the table.

Part of tricycle	Properties this part of the tricycle must have	Material
tyres		
brake		
frame		
seat		
handle to push tricycle		
pushing pole		
screws that join pushing pole to handle		
bag		

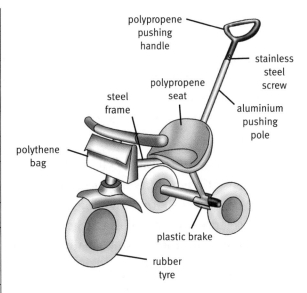

polypropene pushing handle

stainless steel screw

polypropene seat

steel frame

aluminium pushing pole

polythene bag

plastic brake

rubber tyre

2 Highlight the correct word in each pair of **bold** words.

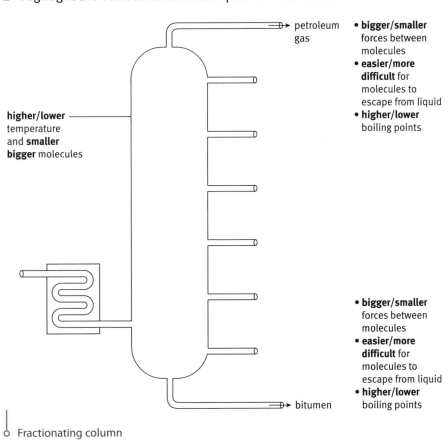

petroleum gas

- **bigger/smaller** forces between molecules
- **easier/more difficult** for molecules to escape from liquid
- **higher/lower** boiling points

higher/lower temperature and **smaller bigger** molecules

- **bigger/smaller** forces between molecules
- **easier/more difficult** for molecules to escape from liquid
- **higher/lower** boiling points

bitumen

Fractionating column

3 Read the article, then complete the tables below it by suggesting reasons for the scientists' observations and actions.

> ### Gold nanoparticles kill cancer
>
> Scientists have recently discovered that gold nanoparticles can kill cancer cells. They removed cancer cells from people with ear cancer, nose cancer, and throat cancer. They used chemicals to make gold nanoparticles enter the cancer cell nuclei.
>
> The nucleus of a cancer cell divides more quickly than the nucleus of a normal cell. The scientists knew that if they could stop a cancer cell from dividing, they could stop the cancer. The scientists observed that, when gold nanoparticles entered a cell nucleus, the cell stopped dividing, and died.
>
> In future, the scientists plan to test whether gold nanoparticles kill cancer cells when they are in the body. The scientists say they need to prevent gold nanoparticles entering healthy cells in the body.

Observation	Suggested reason
Nanometre-sized particles get into the nuclei of cancer cells more easily than normal-sized gold particles.	

Action	Suggested reason
The scientists tried to stop cancer cells dividing.	
The scientists did the tests on cancer cells outside the body.	
The scientists took cancer cells from many people.	
In future, when the scientists do tests on cancer cells inside the body, they will try to prevent gold nanoparticles entering healthy cells.	

4   Draw lines to match each modification to **one** diagram and **one or more** changes in properties. Use each change in property once, more than once, or not at all.

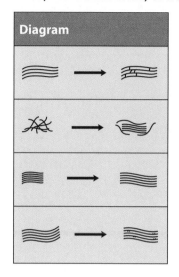

Diagram	Modification	Changes in properties
	increase chain length	stronger
		less flexible
	add plasticiser	softer
		harder
	make cross-links between polymer chains	more flexible
		more dense
	pack molecules neatly together with crystalline regions	less dense

5   Write a C in the boxes next to the equations for combustion reactions.
Write a P in the boxes next to the equations for polymerisation reactions.

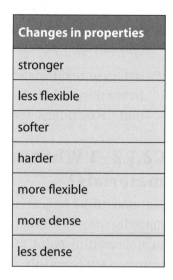

a

b

c

d

Key :

● = carbon atom

○ = hydrogen atom

● = oxygen atom

6   For each equation, draw more oxygen, carbon dioxide, or water molecules in the box so that there are the same number of atoms of each element in the products and reactants.

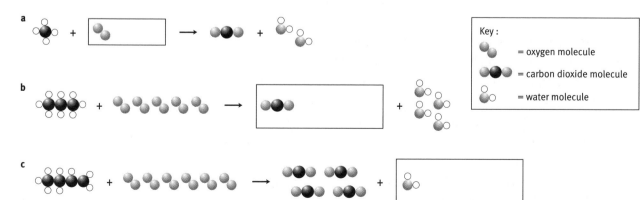

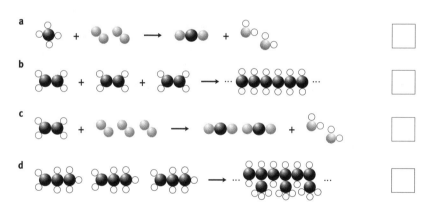

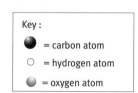

a

b

c

Key :

= oxygen molecule

= carbon dioxide molecule

= water molecule

C2.2.1–2 What are materials made from?

Every material is a chemical or mixture of chemicals, including:
- ceramics, for bricks, wall tiles, and plates
- metals, for vehicles, high-rise buildings, and jewellery
- polymers, for packaging and protective clothing.

We obtain or make materials from:
- living things, for example cotton, wool, leather, and wood
- non-living things, for example limestone and oil.

C2.1.2–3 What are the properties of materials?

Manufacturers look at materials' properties to choose the best material from which to make a product. Important properties include melting point, strength in tension (pulling), strength in compression (squashing), stiffness, hardness, and density.

The effectiveness and **durability** of a product depend on the properties of its materials. A product that is durable lasts for a long time before breaking or rotting.

C2.2.7 What's in crude oil?

Crude oil is a thick, dark-coloured liquid. It is mainly a mixture of **hydrocarbon** molecules of different lengths.

Hydrocarbon molecules are made from hydrogen and carbon only, for example propane and octane.

propane

octane

Key :
● = carbon atom
○ = hydrogen atom

C2.2.9–11 How is crude oil used?

Crude oil is not much use as it is. So oil companies use **fractional distillation** to separate crude oil into **fractions**. A fraction is a mixture of hydrocarbons with similar boiling points.

Fractional distillation happens in a fractionating tower. It works like this:
- Crude oil is heated in a furnace. Its compounds evaporate and become gases.
- The gases enter the tower. As they move up, they cool down. Different fractions condense at different levels:
  - Compounds with small molecules have low boiling points. This is because the forces between the molecules are weak, so only a little energy is needed for them to break out of a liquid and form a gas. These molecules rise to the top of the tower.
  - Compounds with big molecules have higher boiling points. They condense at the bottom of the tower.

Different fractions have different uses, including:
- fuels, e.g. methane, petrol, diesel, and liquefied petroleum gas
- lubricants, e.g. Vaseline and engine oil
- raw materials to make new materials in **chemical synthesis**.

C2.2.3–6, C2.2.8, C2.2.12–13 How are polymers made?

Most of the chemicals obtained from crude oil are used as fuels. Just 4% of crude oil makes **synthetic materials**, such as polymers. Synthetic materials do not occur naturally. They are made in chemical processes from raw materials from the Earth.

Polymers are very long molecules. They form when many small molecules, called **monomers**, join together. This type of chemical reaction is called **polymerisation**.

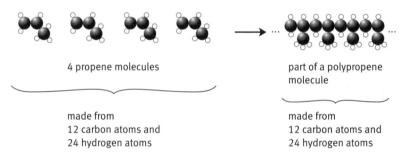

4 ethene molecules   part of a polythene molecule

In polymerisation reactions – as in all chemical reactions – there are the same numbers of atoms of each element in both the reactants and products. The atoms are **rearranged**.

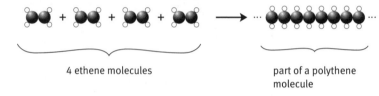

4 propene molecules   part of a polypropene molecule

made from
12 carbon atoms and
24 hydrogen atoms
made from
12 carbon atoms and
24 hydrogen atoms

There are many polymers, all made from different starting materials. Each polymer has unique properties.

Synthetic polymers have replaced natural materials in many products. For example, many ropes are now made with polypropene instead of sisal. Clothes may be made from nylon instead of cotton, because nylon is more durable.

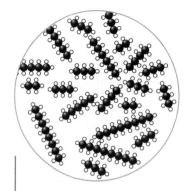

Crude oil is a mixture of hundreds of different hydrocarbons.

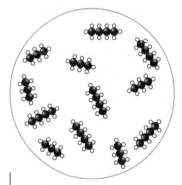

These molecules go right to the top of the fractionating column. This fraction contains some of the shortest hydrocarbons. There are weak forces between the molecules, giving them a low boiling point.

Exam tip

Remember, the bigger the molecule, the higher the boiling point.

C
2

C2.3.1–3 What gives polymers their properties?

The properties of polymers depend on how their molecules are arranged and held together. For example, wax has shorter molecules than polythene. This means that:

- Polythene is stronger, since its long molecules are tangled and difficult to separate.
- Polythene melts at higher temperatures since there are stronger forces between long polythene molecules than between shorter wax molecules.

C2.3.4 What makes polymer properties change?

Method	How properties change	Diagram
making chains longer	• stronger	
adding cross-links	• harder • stronger • less flexible	
adding plasticisers	• softer • more flexible	
increasing crystallinity by lining up polymer molecules	• stronger • denser	

C2.4.1–6 What is nanotechnology?

Nanotechnology is the use and control of tiny structures. **Nanoparticles** are about the same size as some molecules, between about 1 and 100 nanometres (nm) across.

Nanoparticles have different properties compared with larger particles of the same material. This is partly because nanoparticles have a bigger surface area compared with their volume.

Nanoparticles can occur:

- naturally, in sea spray
- by accident, when fuels burn
- by design, for example:
  - silver nanoparticles give fibres antibacterial properties, in medical dressings and socks
  - adding nanoparticles to plastics for sports equipment such as tennis rackets makes them stronger.

Nanoparticles may have harmful effects on health. Some people think these effects should be studied more closely before using nanoparticles more widely.

Use extra paper to answer these questions if you need to.

1 Tick the boxes to show which materials listed below are obtained or made from living things.

a cotton ☐ d wool ☐
b copper ☐ e paper ☐
c limestone ☐ f silk ☐

2 Highlight the statements below that are **true**. Then write corrected versions of the statements that are **false**.

a A synthetic material is one that is obtained from living things.

b A hydrocarbon is a compound made of carbon, hydrogen, and oxygen only.

c In a polymerisation reaction, small molecules join together to make very long molecules.

d Monomers are very long molecules.

e Most crude oil is used for chemical synthesis.

f In a chemical reaction, there are always more atoms of each element in the product than in the reactants.

3 Choose words or phrases from the box to fill in the gaps in the sentences below. The words in the box may be used once, more than once, or not at all.

molecules NM 100 atoms seaspray
10000 fuel combustion products
grains of sand 1 10 nm nM Nm

Nanotechnology involves structures that are about the same size as some _____ . Nanoparticles have diameters of between _____ and _____ nanometres. Nanometre is also written as _____ . Nanoparticles can occur by accident in _____ . They also occur naturally in _____ . Scientists also make nanoparticles.

4 The table shows the properties of two materials used to make dental fillings. Use the data in the table to suggest why many people now prefer polymer fillings, even though amalgam fillings have been used for many more years.

Property	Amalgam	Dental polymer
Conduction of heat	good conductor of heat	poor conductor of heat
Colour	silver	white
Risks	contains mercury, which is poisonous	no known health risks

5 The table shows the properties of two materials: polypropene and sisal.

Property	Polypropene	Sisal
durability	does not rot	rots
colour	can be pigmented any colour	can be dyed any colour
relative strength in tension	1.4	0.8
flexibility	very flexible	very flexible

Ropes for life buoys near rivers used to be made from sisal. Now they are made from polypropene. Select data from the table to suggest two reasons that explain why polypropene is now preferred for making life buoy ropes.

6 The table shows the properties of two polymers, LDPE and HDPE. Use data from the table to explain why HDPE is the better polymer for making garden furniture.

Property	LDPE	HDPE
density (g/cm^3)	0.92	0.95
maximum temperature at which the polymer can be used (°C)	85	120
strength (MPa)	12	31
relative flexibility	flexible	stiff

7 Highlight the correct word or phrase in each pair of **bold** words.

Kerosene and petrol are two **fractions/polymers** obtained by the fractional distillation of crude oil. The hydrocarbon molecules in kerosene are bigger than those in petrol. So the forces of attraction between molecules in kerosene are **greater/smaller** than those in petrol. This means that the hydrocarbons in kerosene have **lower/higher** boiling points than those in petrol. Kerosene is removed from **lower down/higher up** the fractionating tower.

8 Give one reason to explain why nanoparticles of a material may show different properties compared with larger particles of the same materials.

9 Give two examples of how nanoparticles are used.

10 Suggest why some people are concerned about the widespread use of products containing nanoparticles.

11 For each of the changes listed below, describe the effect it has on the properties of a polymer, and explain why it has this effect. Include diagrams to help you explain your answer.

a increased chain length c adding plasticisers
b cross-linking d increasing crystallinity

C2

1 The table shows the properties of some synthetic polymers.

Letter	Name of material	Properties
A	poly(2-hydroxyethylmethacrylate) (PHEMA)	transparent; absorbs water to become flexible and jelly-like
B	acrylic	easily moulded into shape
C	polyethenol	flexible, soluble in water
D	silicone rubber	insoluble in water; very durable

a Study the properties of the materials in the table. Choose the best material to make each of the following items. Write the letter of one material next to each item.

artificial heart valves ☐

hospital laundry bags that dissolve in a washing machine, allowing dirty sheets to be washed without needing to be handled ☐

part of a composite used to make fillings for front teeth ☐

contact lenses ☐ [4]

b Disposable nappies are made from several materials, including:

cellulose   polypropene   polythene

i From the list above, write the name of one material that is obtained from a living thing.

_____ [1]

ii Draw rings round the properties that the outermost layer of a disposable nappy must have.

**non-toxic   hard   flexible
high strength in tension   stiff** [2]

iii Polythene is made when small molecules join together to make very long molecules. Give the name of this process.

_____ [1]

Total [8]

2 A student wants to make a one-person rowing boat to use on a lake.

He investigates three materials.
He does experiments to obtain the data in the first two lines of the table on the next page.

He uses the Internet to collect the data in the last two lines of the table.

Property	Aluminium alloy 5083	ABS steel (an alloy of iron)	Glass-reinforced plastic
Mass (g)	30.5	115.0	20.7
Volume (cm³)	11.5	16.2	14.8
Density (g/cm³)			
Tensile strength (MPa) (the force to pull a material until it breaks)	300	between 400 and 490	varies, but lower than aluminium and steel
Yield strength (MPa) (how much a material bends before it won't spring back into shape)	150	235	low – the material is brittle and shatters easily on collision

a Use the equation below to calculate the density of each material in the table.

Write your answers in the empty boxes in the table.

density = mass ÷ volume [3]

b Use all the data in the table, including your calculated density values, to choose the best material for the rowing boat. Give reasons for your choice.

_____

_____

_____

_____ [4]

Total [7]

3 The table gives data about three building materials.

Material	Compressive strength, in MPa (how much pushing force the material can withstand before it is crushed)	Thermal conductivity, in W/mK (how well the material conducts heat – the bigger the number, the better the material conducts heat)
Limestone	60	1.3
High strength concrete	60	1.7
Wood (oak)	15	0.1

Use only the data in the table to evaluate the advantages and disadvantages of using the three materials listed in the table to build a house.

✎ The quality of written communication will be assessed in your answer to this question.

Write your answer on separate paper or in your exercise book.

Total [6]

4 a The chemicals used to make candle wax and shoe polish are obtained from crude oil.
  - Candle wax is made from hydrocarbon chains that are about 30 carbon atoms long.
  - The waxy ingredient of shoe polish is made from hydrocarbon chains that are about 70 carbon atoms long.

Tick the **two statements that best explain** why shoe polish wax has a higher melting point than candle wax.

The forces between long hydrocarbon molecules are stronger than the forces between short hydrocarbon molecules. ☐

The forces between long hydrocarbon molecules are weaker than the forces between short hydrocarbon molecules. ☐

The stronger the forces between molecules, the smaller the amount of energy needed to separate them. ☐

The stronger the forces between molecules, the more energy is needed to separate them. ☐ [2]

b i PVC is a polymer. It is used to make these items:
  - window frames
  - floor coverings
  - shower curtains

Write the name of the item that has the most plasticiser added to the PVC that it is made from.

_____

Give a reason for your choice.

_____ [2]

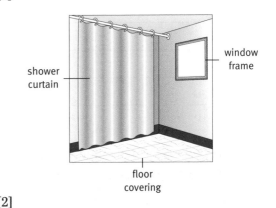

shower curtain

window frame

floor covering

ii Rubber is used to make car tyres and elastic bands.

Predict which has more cross-linking: the rubber in car tyres or the rubber in elastic bands.

_____

Give a reason for your choice.

_____ [2]

Total [6]

1 Decide which statements apply to chlorine, which statements apply to sodium chloride, and which statements apply to sodium hydroxide.

Write the letter of each statement in the correct part of the Venn diagram.

A an element

B a compound

C includes chlorine atoms

D dissolves in water to make an alkaline solution

E obtained by solution mining

F obtained from the sea

G a chemical

H includes sodium atoms

I obtained by the electrolysis of brine

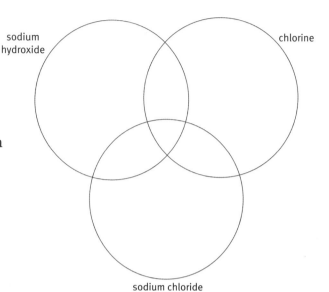

2 Complete the table to summarise what geologists can find out from the clues.

Clue	What geologists can learn about a rock from this clue

Exam tip

Elements are made up of one type of atom. Compounds are made up of atoms of two or more elements, strongly joined together. The properties of a compound are very different to the properties of the elements it is made up of.

C
3

3 Annotate the diagram to explain how salt is obtained by solution mining.

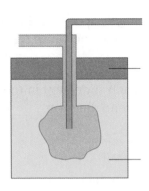

4 Solve the clues to fill in the grid.

1 An alkali reacts with fat to form ...

2 When hydrogen chloride reacts with oxygen, it is ...

3 Potassium ... is an alkali.

4 The raw materials for making an alkali in the industrial revolution were salt, coal, and ...

5 Pure sodium chloride is obtained from the sea and by ... mining.

6 Scientists use ... clues in rocks to track the past movements of continents.

7 The outer layer of the Earth is made up of ... plates.

8 Two gases are obtained by the electrolysis of brine – chlorine and ...

9 The food industry uses salt as a ... and preservative.

10 PVC is made up of long chains of atoms, so it is a ...

11 When added to water, ... kills microorganisms.

12 Geologists get clues about the relative ages of rocks from ...

13 Some synthetic chemicals are dangerous because they do not break down in the environment, so they travel long ...

14 Eating too much salt can raise you blood ...

C3.1.1–5, C3.1.7 How were rocks made?

The outer layer of the Earth is made up of about 12 **tectonic plates**. Convection currents under the plates make them move.

Scientists use magnetic clues in rocks like **magnetite** to track continents' movements. The movements mean that the parts of the ancient continents that now make up Britain have moved over the surface of the Earth. So different rocks, like salt, limestone, and coal, were formed in different climates.

These processes form rocks:
- **Sedimentation** and compression formed Peak District limestone when remains of dead sea animals fell to the bottom of warm seas near the equator.
- **Erosion** of rocks by rivers formed sand, which was deposited in layers to form sandstone.
- **Evaporation** formed rock salt when a sea moved inland, and its water evaporated. The salty water formed originally when salts from rocks dissolved in water flowing over them.
- **Mountain building** pushed coal towards the surface in the Peak District. The coal formed when tree ferns in swamps died, and were compressed and heated.

Chemical industries grew up where resources were available.

C3.1.6 How do we know how rocks were made?

Geologists study sedimentary rocks to find evidence of the conditions when they formed.
- Different animals lived at different times, so their **fossils** tell us about the ages of the rocks they are in
- Comparing **sand grains** in deserts and rivers to sand grains in sandstone tells us what sort of sand formed the sandstone
- The shapes of **ripples** in rocks give clues about whether sandstone was made from river bed sand or desert sand.
- Tiny **shell fragments** in limestone tell us about the conditions when the rock formed.

C3.2.1–5 Where does salt come from?

We use salt (sodium chloride) in food, to treat icy roads, and as a source of chemicals.

Salt for food must be pure. Some of it comes from the sea. The water evaporates, leaving salt behind. Countries with hot, dry climates get lots of salt from the sea, since energy costs are lower.

C 3

Rock salt is a mixture of salt and clay. It is used for de-icing roads, so need not be pure. It is mined from under the ground with big machines.

Salt for the chemical industry must be very pure. In Britain, it is obtained by solution mining. In solution mining, water is pumped into rock salt. Salt dissolves in the water underground. The salt solution is then pumped to the surface, and solid salt is extracted from the solution.

In the past, solution mining made huge underground holes. This caused **subsidence**, and buildings fell into the holes. Now, miners leave pillars in mines to prevent this problem.

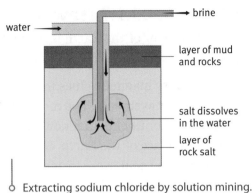

Extracting sodium chloride by solution mining.

C3.2.6–9 Why is salt added to food?

Salt is used by the food industry:
- as a **flavouring**, to improve the flavour of food
- as a **preservative**, to stop food going off.

Eating too much salt can raise your blood pressure. This increases the risk of having a stroke or heart attack.

Government departments do risk assessments on food chemicals like salt. They tell the public about the risks.

C3.3.1, C3.3.3–5 Why are alkalis useful?

Before industrialisation, alkalis were used to:
- neutralise acid soil
- make chemicals to bind dyes to cloth
- convert fats and oils into soap
- make glass.

Alkalis dissolve in water to make solutions with a pH above 7. They react with acids to form **salts**. The **word equation** shows how sodium hydroxide neutralises an acid.

sodium hydroxide + hydrochloric acid $\longrightarrow$ sodium chloride + water

Alkalis include soluble hydroxides and soluble carbonates. Soluble hydroxides react with acids to form a salt and water:

sodium hydroxide + sulfuric acid $\longrightarrow$ sodium sulfate + water

potassium hydroxide + nitric acid $\longrightarrow$ potassium nitrate + water

Soluble carbonates react with acids to make a salt, water, and carbon dioxide:

sodium carbonate + hydrochloric acid $\longrightarrow$ sodium chloride + water + carbon dioxide

Exam tip

Don't confuse sodium hydroxide (an alkali) with sodium chloride (common salt).

C3.3.2, C3.3.6–7 How were alkalis obtained?

Before the industrial revolution, people got alkalis from burnt wood and from stale urine.

In the industrial revolution, there was a shortage of alkali. A scientist invented a new way of making an alkali on a large scale. The raw materials were salt (sodium chloride), limestone (calcium carbonate), and coal.

The process made huge amounts of pollutants, including:
- acidic hydrogen chloride gas
- solid waste, which emitted toxic, smelly hydrogen sulfide gas.

In 1874, Henry Deacon worked out how to use one of the pollutants, hydrogen chloride, to make a useful product, chlorine. Hydrogen chloride is oxidised with oxygen:

> hydrogen chloride + oxygen ⟶ chlorine + water

Hydrogen chloride is a compound. Its properties are different from those of the elements it is made up of, chlorine and hydrogen.

Chlorine is used as a bleach, and to whiten paper and textiles.

C3.3.10–13 Why add chlorine to water?

Sewage-contaminated water may contain microorganisms that cause cholera and typhoid. Chlorine kills the microorganisms. Adding chlorine to water leads to fewer deaths from waterborne diseases.

There may be disadvantages to chlorinating water. When chlorine reacts with **organic matter**, trihalomethanes (**THMs**) may form. Some people think that drinking this water can cause cancer.

C3.3.14–17 What does brine make?

Today, chlorine is made from sodium chloride solution (brine). Passing electricity through brine causes chemical changes. The elements in the sodium chloride (sodium and chlorine) and water (hydrogen and oxygen) are rearranged to make new products:
- chlorine gas – used to treat water, and to make bleach, plastics, and hydrochloric acid
- hydrogen gas – a fuel, and used to make hydrochloric acid
- sodium hydroxide solution – used to make soap, paper, and bleach.

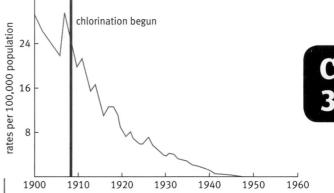

Death rate from typhoid fever in the USA, 1900–1960 (first published in the US Center for Disease Control and Prevention's Summary of Notifiable Diseases 1997).

Exam tip
Watch out! Chlorine becomes chloride in compounds like sodium chloride.

C 3

The electrolysis of brine needs electricity. If the electricity is generated from fossil fuels, much pollution results.

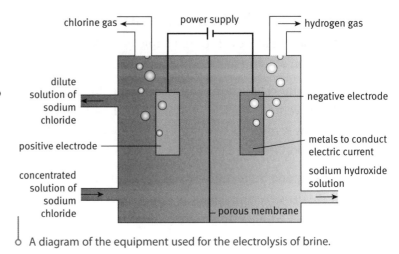

A diagram of the equipment used for the electrolysis of brine.

C3.4.1–2 Are chemicals risky?

Industry makes and uses many **synthetic chemicals**. In big quantities, some may harm health. But there is no evidence that the tiny amounts of these chemicals in human blood are unsafe.

For many chemicals, there are not enough data to judge whether the chemicals are likely to present a risk to health or the environment.

But there are twelve synthetic chemicals that are banned because everyone agrees are harmful, even in tiny amounts. These cause problems because:
*   they do not break down in the environment
*   they move long distances in the air and water
*   they build up in fatty tissues of animals and humans.

C3.4.3–4 What are the dangers of PVC?

PVC is a useful synthetic chemical. It is a **polymer** – its molecules are made up of chains of carbon, hydrogen, and chlorine atoms.

Hard PVC makes window frames and underground water pipes. Softer PVC makes electric wire insulation, and clothing. PVC film makes hospital blood bags and drip bags.

PVC is softened by adding **plasticisers**. Plasticisers have small molecules. They can escape from the plastic and dissolve in liquids in contact with it. There is some evidence linking plasticisers to cancer and infertility. In the EU, some plasticisers have been banned from toys. PVC makers say plasticisers have never harmed anyone.

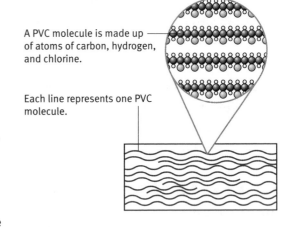

A PVC molecule is made up of atoms of carbon, hydrogen, and chlorine.

Each line represents one PVC molecule.

C3.4.5–6 What is a life cycle assessment?

We can use **life cycle assessments** (LCAs) to analyse the stages of the life of a product. LCAs include assessments of:
*   the use of resources, including water
*   the energy inputs or outputs
*   the environmental impact.

Stage	Environmental impact
materials are made	raw materials to make material energy and water in processing
manufacturers make the product	materials to make product energy and water in manufacture
people use the product	energy to use product (e.g. petrol for a car) energy, water, and chemicals needed to maintain product
people get rid of the product	energy to take product away (fuel for bin lorry) space to store rubbish

Use extra paper to answer these questions if you need to.

1 Write an R next to the raw materials of the first industrial process for making sodium hydroxide (an alkali). Write a P next to the pollutants made in this process.

a coal   d limestone
b hydrogen sulfide   e salt
c hydrogen chloride

2 Draw lines to match each chemical to one or more of its uses.

Chemical
chlorine
sodium hydroxide
hydrogen
sodium chloride

Use
to de-ice roads
to make hydrochloric acid
to make soap
to preserve food
as a fuel
to make bleach

3 Highlight the one correct word or phrase in each pair of **bold** words or phrases.
PVC is a **plasticiser/polymer**. It is made up of **short/long** chains of atoms, including atoms of **carbon/sodium**. Hard PVC is used to make **window frames/wire insulation**. Soft PVC makes **window frames/wire insulation**. PVC is made soft by adding a **plasticiser/polymer**. This has **big/small** molecules, which can escape from the plastic. In the EU, some of these have been banned from **toys/clothes**.

4 Use the words in the box to fill in the gaps. Each word may be used once or more than once.

| materials   energy   discarded |
| chemicals   product |

In a life cycle assessment, the first stage is to look at how the _____ are produced. Then assess the energy and water to make the _____ . Next, look at how much _____ and water are needed to use the product, and whether any _____ are needed to use or maintain it. Finally, consider what happens to the product when it is _____ . Can it be reused or recycled? How much _____ is needed to take it away?

5 Draw lines to match each process to a description of how it helps to form rock.

Process	How it helps form rock
sedimentation	formed sand, which deposited in layers to make sandstone
erosion	pushed coal nearer the surface
evaporation	with compression, formed limestone when dead sea creatures sank
mountain building	formed rock salt when a sea moved inland

6 The table shows the percentage of deaths in American cities from four diseases in 1900 and 1936. Chlorination of drinking water was introduced in many cities between 1900 and 1936.

Cause of death	Percentage of deaths in major cities	
	1900	1936
tuberculosis	11.1	5.3
pneumonia	9.6	9.3
typhoid	2.1	0.1
flu	0.7	1.3

a Which diseases caused a smaller percentage of deaths in 1936 than in 1900?
b From the data, can you conclude that adding chlorine to drinking water *caused* the reduction in the percentages of deaths from these diseases? Explain your decision.

7 Twelve synthetic chemicals, including DDT, are banned because everyone agrees they are harmful, even in small amounts. Give three reasons why they cause problems.

H 8 Name the products of the following reactions.
a sodium hydroxide and nitric acid
b potassium hydroxide and sulfuric acid
c potassium carbonate and hydrochloric acid
d sodium carbonate and sulfuric acid
e potassium hydroxide and hydrochloric acid
f sodium carbonate and nitric acid

9 Write word equations for each of the reactions in question 9.

C3

1 This question is about the life cycle assessment (LCA) of a computer.

a Draw lines to link each activity to a stage in the life cycle of a computer.

Each stage in the life cycle may be linked to zero, one, or two activities.

Activity
putting together the computer components in its plastic case
recycling the computer components
extracting oil from wells beneath the sea
dismantling the computer
making plastics from oil

Stage in life cycle
Materials are produced.
Manufacturers make the computer.
People use the computer.
People throw away the computer.

[4]

b The data show the mass of carbon dioxide gas emissions during the manufacture of the different parts of a computer, and during its use.

Activity	Mass of carbon dioxide emissions (kg)
manufacture of flat screen	185
manufacture of electronic components	69
manufacture of chemicals used in computer	48
manufacture of plastic casing	17
manufacture of silicon wafers	15
manufacture of circuit boards	11
using the computer for one year	940

[Data from UNEP]

i Calculate the total mass of carbon dioxide emissions for the manufacture of all the computer components listed in the table **and** for using the computer for one year.

Answer = _____ kg [2]

ii Calculate the total mass of carbon dioxide emissions for the manufacture of all the computer components listed in the table **and** for using the computer for five years.

Answer = _____ kg [2]

iii Give **two** reasons to explain why it is less damaging to the environment to use a computer for five years, than for one year, before replacing it with a new one.

_____

_____ [2]

c This part of the question is about recycling computers.

i It has been estimated that one tonne of waste from electronic products (including computers) contains about 225 g of the precious metals gold, silver, and palladium.

On average, there is about 170 g of these metals in one tonne of rock that contains these metals.

Describe **one benefit to the environment**, and **one benefit to people**, of using recycled precious metals in new computers, rather than getting the metals out of rocks from the ground.

_____

_____ [2]

ii Read the article opposite, and then answer the question beneath it.

Identify one risk of dismantling old computers. Suggest why people do dismantle computers at the tip, even though there are risks in so doing.

_____

_____

_____ [2]

Total [14]

2 Describe three different methods by which alkalis have been manufactured in different stages of history.

Identify the advantages and disadvantages of each method.

✎ The quality of written communication will be assessed in your answer to this question.

Write your answer on separate paper or in your exercise book.

Total [6]

3 The box gives the maximum daily amount of salt for children of different ages, recommended by the British Government.

a Explain why the British Government gives recommendations for salt intake.

_____

_____ [1]

Illegal computer dumping risks lives

Some computers that have been sent for recycling are in fact shipped to poorer countries. Here they are dumped in huge tips.

The waste contains hazardous materials, such as the beryllium used in circuit boards. Beryllium dust is toxic to humans. The element and its compounds can cause cancer.

At the tips, people dismantle the computers and remove gold and other valuable metals.

C 3

1 to 3 years old: 2 g of salt/day
4 to 6 years old: 3 g of salt/day
7 to 10 years old: 5 g of salt/day
11 years and older: 6 g of salt/day

b The table gives the amounts of salt in two products, made by the same company, to sell in different countries.

Country	Mass of salt in food (g)	
	100 g of bran cereal	One bacon double cheeseburger
Canada	2.15	–
USA	0.65	–
Brazil	–	3.2
UK	1.13	2.1

i In which of the countries in the table is there the greatest mass of salt in bran cereal?

_____ [1]

ii Suggest two reasons to explain why the same manufacturer adds different amounts of salt to the same food in different countries.

_____

_____ [2]

iii Pedro is 16. He lives in Brazil.
In one day, he eats one bacon double cheeseburger at lunchtime, and one in the evening.
Calculate the total mass of salt in the burgers.

_____

_____ [1]

iv Compare the mass of salt in Pedro's burgers with the recommended daily maximum amount of salt for a 16-year-old.

What advice would you give Pedro?
Give reasons to support your answer.

_____

_____ [2]

Total [7]

> **Exam tip**
>
> When using data from tables, make sure you check the units and amounts given in the column headings.

Going for the highest grades

4 Predict the names of the products of the following reactions.

a copper carbonate with hydrochloric acid

_____ [2]

b sodium hydroxide with nitric acid

_____ [2]

c potassium hydroxide with sulfuric acid

_____ [2]

Total [6]

1 **a** Fill in column B of the table below. Choose words from the box.

| harmful | toxic | explosive | corrosive |
| oxidising | highly flammable |

b In column C, write down one safety precaution you must take when using a chemical that displays the hazard symbol. (Assume you are already wearing eye protection.)

A Hazard symbol	B Meaning of symbol	C Safety precaution
(corrosive symbol)		
(toxic symbol)		
(oxidising symbol)		

2 Write the symbol of each element in the box below its proton number. You can find these in the periodic table on page 8.

Proton number	3	26	53	16	9	92	7	16	8	53	16	75	23	53	14	8	7
Symbol	Li					U											

Now crack the code. What does the sentence say?

_____

3 Join the dots to make a picture.

Start at the element with the lowest relative atomic mass. Join this to the element with the next lowest relative atomic mass, and so on.

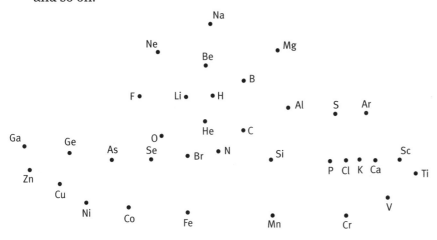

4   Write the symbol of each element in the box below its relative atomic mass. You can find this information in the periodic table on page 8.

Relative atomic mass	32	4	127	32	9	197	48	19	238	7	232	127	14	39
Symbol									U		Th			

Now crack the code. What does the sentence say?

_____

5   On the periodic table:
- colour in red the group that includes the element calcium
- colour in blue the period that includes the element phosphorus
- colour in pencil all the non-metals
- circle in red **three** elements that form ions with a charge of +1
- circle in blue **three** elements that form ions with a charge of −1.

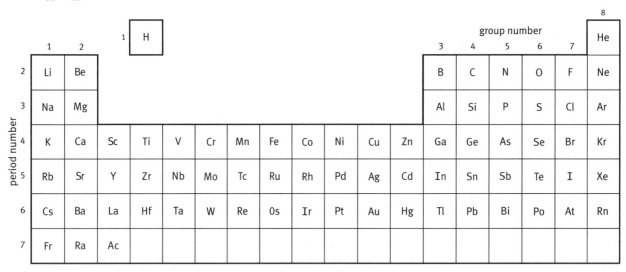

6   Draw crosses on the circles to show how the electrons are arranged in atoms of these elements. (Hint: not all of the shells will contain electrons.)

lithium

beryllium

carbon

fluorine

sodium

phosphorus

chlorine

argon

C4.1.2, C4.1.6–9, C4.2.5 What is the periodic table?

The periodic table is shown on page 8. In the periodic table:
- The elements are arranged in order of **proton number**.
- There are repeating patterns in the element's properties.
- The vertical columns are called **groups**.
- The elements in a group have similar properties.
- The horizontal rows are called **periods**.
- The elements to the left of a stepped line between aluminium and silicon, germanium and arsenic, and so on, are metals. The elements to the right of this stepped line are non-metals.

C4.1.3–5 How was the periodic table created?

In the early 1800s, Döbereiner noticed that there were several groups of three elements with similar properties.

Newlands arranged the elements in order of relative atomic mass. Every eighth element had similar properties. His pattern only worked for the first 16 elements.

Mendeleev showed there were patterns in the properties of all the elements when arranged in relative atomic mass order. He made a sensible pattern by leaving gaps for missing elements. He predicted the missing elements' properties. Later, scientists discovered these elements.

C4.1.18 What do hazard symbols mean?

Symbol	Meaning	Safety precautions Wear eye protection and . . .
	Toxic – can cause death if absorbed by skin, swallowed or breathed in	• wear gloves • work in fume cupboard or wear mask over mouth and nose
	Harmful – like toxic substances, but less dangerous	• wash off spills quickly • use in a well-ventilated room
	Corrosive – attacks surfaces and living tissue, like eyes or skin	• wear gloves
	Explosive	• avoid situations that could initiate an explosion
	Highly flammable – catches fire easily	• keep away from flames, sparks and oxidizing chemicals
	Oxidising – provides oxygen so other chemicals burn more fiercely	• keep away from flammable chemicals

C
4

C4.1.11–19, C4.1.28 (part), 29, 30 What are the patterns in the properties of the elements?

Group 1: the alkali metals

The alkali metals include the elements lithium (Li), sodium (Na), and potassium (K). The alkali metals:

- have low densities, so they float on water
- have **low melting** and **boiling points**
- are **shiny** when freshly cut
- quickly **tarnish** in damp air because they react with oxygen.

Alkali metals react with water to make **hydrogen** and an **alkaline solution**.

For example:

$$\text{sodium} + \text{water} \longrightarrow \text{hydrogen} + \text{sodium hydroxide}$$
$$2Na(s) + 2H_2O(l) \longrightarrow H_2(g) + 2NaOH(aq)$$

The symbol equation shows that two atoms of sodium react with two water molecules to make one hydrogen molecule and two formula units of sodium hydroxide.

Going down the group, the reactions get more vigorous.

Alkali metals also react vigorously with **chlorine gas** to make chlorides. The chlorides are **colourless crystalline solids**. Again, the reactions get more vigorous going down the group.

For example:

$$\text{lithium} + \text{chlorine} \longrightarrow \text{lithium chloride}$$
$$2Li(s) + Cl_2(g) \longrightarrow 2LiCl(s)$$

C4.1.20–27, C4.1.28 (Part) Group 7: the halogens

Name and symbol	Formula	State at room temperature	Colour
chlorine, Cl	Cl_2	gas	pale green
bromine, Br	Br_2	liquid	deep red liquid with red-brown vapour
iodine, I	I_2	solid	grey solid with purple vapour

Going down the group, **melting point** and **boiling point** increase.

Halogen molecules are **diatomic** – they are made from two atoms joined together. For example, the formula of bromine is Br_2.

C 4

Going **down** group 7, the elements become **less reactive**. For example:

- Hot iron glows brightly in chlorine gas. It glows less brightly in bromine, and hardly at all in iodine.

  iron + chlorine $\longrightarrow$ iron chloride

- James adds pale green chlorine solution to a colourless solution of sodium bromide. Red bromine solution forms. Chlorine is more reactive than bromine, so it displaces bromine from its salt in a **displacement** reaction.

  chlorine + sodium bromide $\longrightarrow$ sodium chloride + bromine

C4.1.1, C4.2.1–4, C4.2.10–14 Explaining patterns in the properties of elements

Particle	Relative mass	Relative charge
proton	1	+1
neutron	1	none
electron	negligible	−1

Atomic structure

An atom has a tiny central nucleus made of **protons** and **neutrons**. Around the nucleus are **electrons**.

All atoms of the same element have the same number of protons. For example, every sodium atom has 11 protons. The **proton number** of sodium is 11.

The number of electrons in an atom is the same as the number of protons. Electrons are arranged in **shells**. Each electron shell fills from left to right across a period.

> **Exam tip**
>
> Practise using the periodic table to work out the number of protons, neutrons, and electrons in atoms.

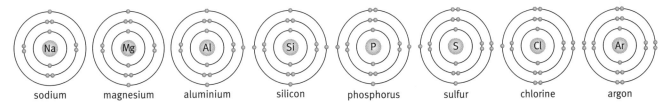

| sodium | magnesium | aluminium | silicon | phosphorus | sulfur | chlorine | argon |

You can also represent the atomic structure of sodium as 2.8.1.

H Every group 1 element has one electron in its outer shell. An element's chemical properties depend on its electron arrangement.
So group 1 elements have similar chemical reactions.

C4.2.6–9 Elements have distinctive flame colours

If you hold any lithium compound in a Bunsen flame at the end of a platinum wire, you see a red flame. The compounds of other elements make different colours.

When the light from the flame goes through a prism, it makes a **line spectrum**.

Every element has a different spectrum. Chemists have studied these spectra and so discovered new elements, for example helium.

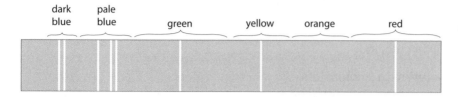

dark blue | pale blue | green | yellow | orange | red

This is the spectrum for helium. The lines are coloured. The series of colours are different for each element.

C4.3.1–3, C4.3.6–9 What are ions and ionic compounds?

If you melt an ionic compound containing metal and non-metal ions, it conducts electricity. Charged particles called **ions** carry the current.

An ion is an atom or group of atoms that has gained or lost electrons. So it has an overall charge.

An atom of sodium has 11 positively charged protons in its nucleus. It has 11 negatively charged electrons. A sodium atom loses one electron to become an ion. A sodium ion has 11 protons and 10 electrons. Its overall charge is +1. Its formula is Na^+.

A chlorine atom has 17 positively charged protons in its nucleus. It has 17 negatively charged electrons. A chlorine atom gains one electron to become an ion. A chloride ion has 17 protons and 18 electrons. Its overall charge is −1. Its formula is Cl^-.

Sodium chloride is a compound that is made from ions. It is **ionic**. Every compound of a group 1 metal with a group 7 metal is ionic.

In solid ionic compounds, the ions are arranged in a regular lattice. So solid ionic compounds form crystals.

When ionic crystals melt or dissolve in water, the ions are free to move independently. So ionic compounds conduct electricity when liquid or in solution.

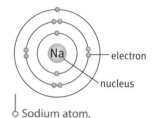

Sodium atom.

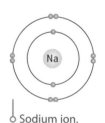

Sodium ion.

Chlorine atom.

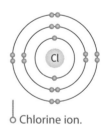

Chlorine ion.

⊕ C4.3.4–5 Ion calculations

The formula of sodium chloride (common salt) is NaCl. There is one sodium ion for every chloride ion. The total charge on the ions in the formula is zero. So sodium chloride, like all compounds, is electrically neutral.

	Sodium ion	Chloride ion	Sodium chloride
Charges	+1	−1	$(+1) + (−1) = 0$

C 4

Working out the formulae of an ionic compound

What is the formula of potassium oxide?

The charge on a potassium ion is $+1$ (K^+).

The charge on an oxide ion is -2 (O^{2-}).

The total charge on the ions in the formula must equal zero.

So potassium oxide has two K^+ ions for every one O^{2-} ion.

So the formula of potassium oxide is K_2O.

Working out the charge on an ion

The formula of calcium bromide is $CaBr_2$.

The charge on one bromide ion is –1.

What is the charge on the calcium ion?

The total charge on the two bromide ions is $-1 \times 2 = -2$.

The total charge on the ions in the formula must be zero (neutral).

So the charge on the calcium ion is $+2$.

C4.1.31–33 Balancing equations

Balance the equation $HCl + MgO \longrightarrow MgCl_2 + H_2O$

- Count the number of hydrogen atoms on each side of the arrow. There are one on the left and two on the right. Write a big 2 to the left of HCl:

$$2HCl + MgO \longrightarrow MgCl_2 + H_2O$$

  Now there are two hydrogen atoms on each side.

- Count the number of chlorine atoms on each side. The big 2 to the left of HCl means that there are two on the left. There are also two on the right of the arrow (in $MgCl_2$). The number of chlorine atoms is balanced.
- Count the number of magnesium atoms on each side of the arrow. There is one on the left and one on the right.
- Count the number of oxygen atoms on each side of the arrow. There is one on the left and one on the right.
- Add state symbols.

So the balanced equation is:

$2HCl(aq) + MgO(s) \longrightarrow MgCl_2(aq) + H_2O(l)$

Exam tip

Never change the formula of a compound or element to balance an equation.

Use extra paper to answer these questions if you need to.

1. Choose words from the box to fill in the gaps in the sentences below. The words may be used once, more than once, or not at all.

shells	electrons	nucleus
neutrons	7	protons	2

Atoms have a small central _____. This is made of protons and _____. Electrons are arranged in _____ round the nucleus. In a neutral atom, the number of _____ is equal to the number of protons. The way an element reacts depends on how its _____ are arranged.

Chlorine, bromine, and iodine are in group _____ of the periodic table. They all have _____ electrons in their outer shell.

2. For the sentences below, write **1** next to each sentence that is true for **group 1**. Write **7** next to each sentence that is true for **group 7**. Write **B** next to each sentence that is true for **both group 1 and group 7**. Use the data in the table to help you.

Element	Boiling point (°C)	Density (g/cm³)
lithium	1342	0.53
sodium	883	0.97
potassium	766	0.86
chlorine	−34	1.56
bromine	58	3.1
iodine	184	4.9

a. Going down the group, boiling point increases.
b. Going up this group, the elements get more reactive.
c. Going up this group, proton number decreases.
d. Atoms of the elements in this group form diatomic molecules.
e. The elements in this group tarnish quickly in damp air.
f. Going down this group, density increases.

3. Complete the word equations.
a. sodium + water $\longrightarrow$ sodium hydroxide + _____
b. potassium + chlorine $\longrightarrow$ _____ _____
c. hydrogen + iodine $\longrightarrow$ _____ _____
d. lithium + _____ $\longrightarrow$ lithium hydroxide + _____
e. sodium + _____ $\longrightarrow$ _____ chloride
f. lithium + _____ $\longrightarrow$ lithium bromide

4. Use the periodic table to work out the number of protons, neutrons, and electrons in atoms of the elements below.
a. sodium   b. phosphorus   c. aluminium
d. vanadium   e. yttrium

5. Fill in the empty boxes.

Name	Formula
water	
hydrogen gas	
	KCl
sodium hydroxide	
	I_2
chlorine gas	
potassium bromide	

6. Decide which of the following pairs of solutions will react together in displacement reactions. Then write word equations for the pairs of solutions that react.
a. chlorine and sodium bromide
b. iodine and potassium bromide
c. bromine and sodium iodide
d. bromine and sodium chloride
e. chlorine and potassium iodide

7. Look at the symbol equation below. It summarises the reaction of iron with chlorine to make iron chloride.
$$2Fe + 3Cl_2 \longrightarrow 2FeCl_3$$
a. How many atoms of iron are shown on the reactant side of the equation?
b. How many molecules of chlorine are shown on the left of the equation?
c. How many atoms of chlorine are shown on the left of the equation?
d. How many formulas of product are made?
e. How many atoms of each element are in one formula of the product?

H 8. Copy and balance the equations. Then add state symbols to show the states at room temperature and pressure.
a. $K + H_2O \longrightarrow KOH + H_2$
b. **Na + Cl₂ $\longrightarrow$ NaCl**
c. $Li + H_2O \longrightarrow LiOH + H_2$
d. $Cl_2 + K \longrightarrow KCl$
e. **Fe + Cl₂ $\longrightarrow$ FeCl₃**

9. Use the information in the table to work out the formulae of the compounds below.

Positive ions	Negative ions
Na^+	Cl^-
K^+	Br^-
Mg^{2+}	O^{2-}
Ca^{2+}	S^{2-}

a. sodium bromide
b. potassium chloride
c. magnesium sulfide
d. potassium oxide

10. Use the information in the table from question 9 to help you answer the questions below.
a. The formula of strontium oxide is SrO. What is the formula of the strontium ion?
b. The formula of beryllium chloride is $BeCl_2$. What is the formula of the beryllium ion?

C 4

1   This question is about strontium and calcium.
    Strontium and calcium are in group 2 of the periodic table.

a   Strontium is present in foods such as cabbage and onions.
    Use the periodic table to suggest why, in the body,
    strontium is absorbed in a similar way to calcium.

    _____

    _____ [2]

b   i   A calcium atom has 20 electrons.
        Finish writing its electronic structure below.

        2.8. _____ [1]

    ii Predict the number of electrons in the outer shell of
        a strontium atom.

        _____ [1]

c   The table shows the properties of some group 2 elements.
    Of the elements shown, calcium is nearest the top of the
    group, and barium is nearest the bottom of the group.

    i   Predict the melting point of strontium.

        _____ °C   [1]

    ii Select data from the table to describe the trend in
        reactivity of the group 2 elements, as you go down
        the group.

        _____

        _____ [1]

d   Strontium reacts with water to make strontium
    hydroxide and hydrogen gas.

    Write a word equation for the reaction of strontium
    with water.

    _____ [2]

e   Use the balanced symbol equation for the reaction of
    strontium with water to answer the questions below.

    $$Sr + 2H_2O \longrightarrow Sr(OH)_2 + H_2$$

    i   How many atoms of strontium are shown on the
        left of the equation?

        _____ [1]

    ii How many molecules of water are shown on the
        left of the equation?

        _____ [1]

    iii Give the total number of hydrogen atoms shown
        in the product side of the equation.

        _____ [2]

                                        Total [12]

> **Exam tip**
>
> If the question asks you to
> select data, you don't need
> to use all the data. Just refer
> to the data that helps you to
> answer the question.

Name of element	Melting point (°C)	Reaction with water
calcium	850	reacts quite fast with cold water to make calcium hydroxide and hydrogen gas
strontium		reacts very fast with cold water to make strontium hydroxide and hydrogen gas
barium	714	reacts very, very fast with cold water to make barium hydroxide and hydrogen gas

2 Ben uses a science data book to find out the melting points of the hydroxides of some group 1 metals.

He plots the melting points on a bar chart.

a Describe the trend shown by the bar chart.

_____ [1]

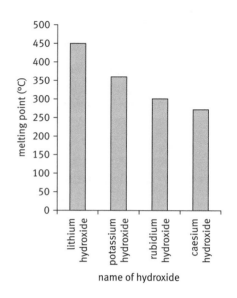

b Ben looks up data about the size of group 1 metal ions.

He writes the data in a table.

Group 1 metal ion	Size (radius) of ion (nm)
lithium ion	0.074
sodium ion	0.102
potassium ion	0.138
rubidium ion	0.149
caesium ion	0.170

Ben develops the explanation below to account for the data in the bar chart.

As the sizes of group 1 metal ions increase, the melting points of their hydroxide compounds decrease.

Ben predicts that the melting point of sodium hydroxide is between 360 °C and 450 °C.

He looks up its melting point in a data book. Its melting point is 319 °C.

Which statement below is correct? Put a tick (✓) in the **one** correct box.

Statement	Tick (✓)
The prediction is incorrect. This proves that the explanation is wrong.	
The prediction is incorrect. This decreases confidence in the explanation.	
The prediction is correct. This increases confidence in the explanation.	
The prediction is correct. This proves that the explanation is correct.	

_____ [1]

c Ben looks up some more data. He writes them in the tables below.

He develops the explanation below.

As the sizes of group 1 and group 2 metal ions increase, the melting points of their chloride compounds decrease.

Name of ion of group 1 metal	Size (radius) of ion (nm)	Melting point of chloride (°C)
lithium	0.074	605
sodium	0.102	801
potassium	0.138	770
rubidium	0.149	718
caesium	0.170	645

Name of ion of group 2 metal	Size (radius) of ion (nm)	melting point of chloride (°C)
beryllium	0.027	405
magnesium	0.072	714
calcium	0.100	782
strontium	0.113	875
barium	0.136	963

C4

Evaluate how well the data in the tables support Ben's explanation.

The quality of written communication will be assessed in your answer to this question.

✎ Write your answer on separate paper or in your exercise book.

_____

Total [8]

3 This question is about sodium fluoride.

a **i** Sodium fluoride consists of sodium and fluoride ions.

The following table shows information about sodium and fluorine atoms and ions.

Complete the table by filling in the empty boxes.

	Number of protons in atom and ion	Number of electrons in atom	Number of electrons in ion	Formula of ion
Sodium	11		10	
Fluorine / fluoride		9		F⁻

[2]

ii Complete the diagram below to show the arrangement of electrons in an ion of sodium. [2]

b Describe what happens to the ions when sodium fluoride dissolves in water.

_____

_____ [1]

c   A solution of sodium fluoride in pure water conducts electricity.

Use ideas about ions to explain why the solution can conduct electricity.

_____

_____ [2]

Total [7]

4   An atom of an element has the electronic structure below:
2.8.8.1

a   **i**   Calculate the number of electrons in an atom of the element.

_____ [1]

ii   Give the number of protons in an atom of the element.

_____ [1]

b   Use your answers to part a and the periodic table to give the name of the element that has the electronic structure 2.8.8.1.

_____ [1]

Total [3]

Going for the highest grades

5   In solution, chlorine solution displaces bromine from sodium bromide solution.

a   Use the periodic table to help you work out the formula of sodium bromide. Write its formula below.

_____ [1]

b   Write a balanced symbol equation, including state symbols, for the displacement reaction of chlorine with sodium bromide solution.

_____ [3]

Total [4]

Exam tip
Never change the formulae to balance an equation.

1 Look at the diagram below and complete the boxes like this:

- Write the letters A, B, and C in the correct small squares.

    A a mixture of elements and compounds that have small molecules

    B a mixture of water and ionic compounds

    C a mixture of minerals

- Write the names of up to four elements or compounds in each box. Choose from this list:

oxygen	sodium chloride	silicon dioxide
argon	water	potassium bromide	carbon dioxide
nitrogen	magnesium chloride	aluminium oxide

- Write the symbols for all the substances in the atmosphere box.

1 Atmosphere

2 Hydrosphere

3 Lithosphere

2 Make up 12 sentences using the phrases in the table.

Each sentence must include a phrase from each column.

Write your answers in the grid at the bottom.

For example, the sentence:

Carbon dioxide... never conducts electricity... because...
no ions or electrons can move freely to carry the current.

is

A	i	10

A Carbon dioxide	a has a high melting point	because	1 the forces of attraction between the molecules are weak.
	b has a low melting point		2 the atoms are held by strong covalent bonds in a giant structure.
	c has a high boiling point		3 it takes a lot of energy to separate its ions to make a gas.
B Silicon dioxide	d has a low boiling point		4 there are strong attractive forces between its oppositely charged ions.
	e conducts electricity when it is liquid or dissolved in water		5 there are strong bonds between its positive ions and a 'sea' of negative electrons.
C Sodium chloride			6 its ions are free to move independently.
	f makes crystals		7 its layers of ions can move over each other.
	g is malleable		8 its oppositely charged ions are held together in a three-dimensional pattern.
	h is very hard		9 much energy is needed to break the strong bonds between the atoms.
D Copper	i never conducts electricity		10 no ions or electrons can move freely to carry the current.

A	i	10

C5.1.1–11 What is the atmosphere?

Dry air is a mixture of gases. Our atmosphere is made up of 78% nitrogen, N_2; 21% oxygen, O_2; 1% argon, Ar; 0.04% carbon dioxide, CO_2.

The chemicals of the air consist of atoms and small molecules. There are weak forces of attraction between the molecules. So these chemicals have low melting and boiling points.

Most non-metal elements, and compounds of non-metals, are molecular, with low melting and boiling points.

The atoms in a molecule are joined together by strong **covalent bonds**. A covalent bond is one or more pairs of electrons shared between atoms.

Covalent bonding arises from electrostatic forces of attraction between the shared electrons and the nuclei of the atoms.

Pure molecular compounds cannot conduct electricity. This is because their molecules are not charged.

$N \equiv N$
nitrogen molecule

$O = O$
oxygen molecule

Ar
argon atom

$O = C = O$
carbon dioxide molecule

C5.2.1–6 What are ionic compounds?

Seawater is a mixture of water and dissolved **ionic compounds,** called salts.

Ionic compounds are made up of positive and negative ions. In ionic solids, the ions are arranged in a **giant three-dimensional lattice** to make crystals. There are very strong attractive forces between the oppositely charged ions. This is ionic bonding.

Ionic compounds have high melting and boiling melting points because much energy is needed to break down an ionic structure.

A solid ionic compound cannot conduct electricity. But when an ionic compound melts, or dissolves in water, its ions are free to move independently. It can now conduct electricity.

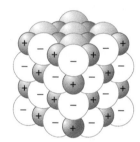
An ionic solid.

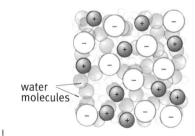
water molecules
A solution of an ionic compound in water.

C5.2.8–15 How can we identify ions?

There are simple tests to identify ions in compounds. They work because each ion, and each compound, has its own properties.

If you mix certain solutions of ionic compounds, you make an insoluble compound. This forms as a **precipitate**. For example:

- Adding an alkali to a solution that contains Cu^{2+} ions makes a blue precipitate of copper hydroxide:

$$CuCl_2(aq) + 2NaOH(aq) \longrightarrow 2NaCl(aq) + Cu(OH)_2(s)$$

You can summarise the equation above with an ionic equation, which shows only the ions that take part in the reaction:

$$Cu^{2+}(aq) + 2OH^-(aq) \longrightarrow Cu(OH)_2(s)$$

C 5

Solutions containing some other metal ions also react with alkalis in solution to make precipitates.

- Adding acidified silver nitrate solution to a solution of Cl^- ions makes a white precipitate of silver chloride:

$$AgNO_3(aq) + NaCl(aq) \longrightarrow AgCl(s) + NaNO_3(aq)$$

$$Ag^+(aq) + Cl^-(aq) \longrightarrow AgCl(s)$$

There are tests to identify other negative ions.

H You can use solubility data to predict chemicals that will precipitate on mixing solutions. A compound with a low solubility will form as a precipitate. If both products are soluble, no precipitate will form.

C5.3.1–8 What are giant structures?

The **lithosphere** is the Earth's rigid outer layer. It consists of the crust and part of the mantle. It is a mixture of **minerals**. Minerals are compounds that occur naturally.

Silicon, oxygen, and aluminium are the most common elements in the Earth's crust. Much of the silicon and oxygen exists as a compound, silicon dioxide. Solid silicon dioxide has a **giant structure** of atoms. Its atoms are held together in a huge lattice by strong covalent bonds.

Diamond and graphite are also minerals. They both consist of carbon atoms arranged in giant structures.

In diamond, each carbon atom is joined to four other carbon atoms by strong covalent bonds. The four bonds are arranged in three dimensions around each carbon atom.

In graphite, each carbon atom is joined to three others by strong covalent bonds. The three bonds are arranged around each carbon atom, making sheets of hexagons. Between the sheets, or layers, are free electrons. These help to stick the layers together.

The structures of diamond, graphite, and silicon dioxide explain their properties and uses.

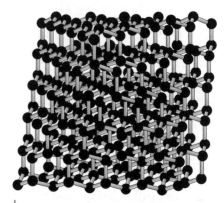

A model of the structure of diamond.

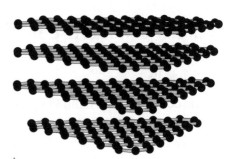

A model of the structure of graphite.

	Graphite	Diamond	Silicon dioxide
Melting and boiling points	Very high because much energy is needed to break the strong covalent bonds in the giant structure of atoms.		
Solubility in water	Insoluble because much energy is needed to break the strong covalent bonds.		
Hardness	Soft because the forces between the layers are weak. A good lubricant.	Very hard because much energy is needed to break the strong covalent bonds between the surface atoms. Diamond is useful for cutting tools and drill tips.	
Electrical conductivity	Good because the electrons between the layers are free to move.	Do not conduct electricity because there are no charged particles free to move.	

C5.4.23–25 Why are metals useful?

Metals have many uses. Their uses depend on their properties. Metals have high melting points, and they are:

- **malleable** – they bend without breaking
- strong
- good electrical conductors.

Solid metals have giant crystalline structures. Strong **metallic bonds** hold the atoms together. This explains why metals are strong and why they have high melting points.

(H) A metal crystal is made up of positive metal ions arranged in layers. When you bend a metal, the layers of ions slide over each other. The ions are held together by a sea of electrons that are free to move. When a metal wire conducts electricity, electrons drift from one end towards the other.

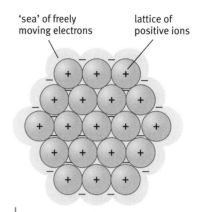

'sea' of freely moving electrons   lattice of positive ions

A model of metallic bonding.

C5.4.1–5, 5.4.11–22 How are metals extracted? (H)

In the lithosphere, most metals are joined to other elements in **minerals**. Rocks that contain useful minerals are called **ores**. Copper is extracted from the mineral copper iron sulfide ($CuFeS_2$). The ore that contains this mineral is copper pyrites.

Ores contain different amounts of minerals. Often, a huge amount of ore contains only a tiny mass of a useful mineral.

The method used to extract a metal from its ore depends on the metal's reactivity.

Extracting metals by heating with carbon

Iron, copper, and zinc are extracted from their oxides by heating with carbon. For example:

$$\text{zinc oxide} + \text{carbon} \longrightarrow \text{zinc} + \text{carbon monoxide}$$
$$ZnO(s) + C(s) \longrightarrow Zn(l) + CO(g)$$

Zinc oxide loses oxygen. It is **reduced**. Carbon gains oxygen. It is **oxidised**.

Extracting metals by electrolysis

Reactive metals, like aluminium, are joined very strongly to other elements in minerals. They cannot be extracted by heating with carbon. So they are extracted by **electrolysis**.

Aluminium oxide is an ionic compound. When it melts, its ions can move independently. So liquid aluminium oxide conducts electricity. It is an **electrolyte**. Electrolytes break down, or **decompose**, when an electric current passes through them. This is **electrolysis**.

C 5

(H) You can calculate the mass of a metal in a mineral. For example:

What is the mass of aluminium in 100 kg of aluminium oxide, Al_2O_3?

- Use the periodic table to find out the relative atomic masses of the elements in the mineral:
  Al = 27 and O = 16
- Calculate the mineral's relative formula mass:
  $(27 \times 2) + (16 \times 3) = 102$
- Calculate the relative mass of the metal in the formula:
  $27 \times 2 = 54$
- Calculate the mass of metal in 1 kg of the mineral:
  $54 \div 102$ kg = 0.53 kg
- Multiply by 100 to find the mass of metal in 100 kg of the mineral:
  0.53 kg $\times$ 100 = 53 kg

To extract aluminium from aluminium oxide:

- Melt aluminium oxide. Pour it into the equipment shown on the right:
- Pass an electric current through the electrolyte.
  - Aluminium (a metal) forms at the negative electrode.
    (H) Al^{3+} ions gain electrons from the electrode to make neutral aluminium atoms:
    $$Al^{3+} + 3e^- \longrightarrow Al$$
  - Oxygen (a non-metal) forms at the positive electrode.
    (H) O^{2-} ions give electrons to the positive electrode to make oxygen atoms:
    $$O^{2-} \longrightarrow O + 2e^-$$
    Oxygen atoms then join together to make oxygen molecules:
    $$O + O \longrightarrow O_2$$

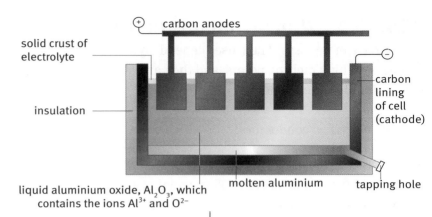

carbon anodes

solid crust of electrolyte

insulation

carbon lining of cell (cathode)

liquid aluminium oxide, Al$_2$O$_3$, which contains the ions Al^{3+} and O^{2-}

molten aluminium

tapping hole

Electrolysis cell for the extraction of aluminium. You do not need to learn the details of this for the exam.

C5.4.26 What are the environmental impacts of using metals?

Use extra paper to answer these questions if you need to.

1 Use the data in the table to answer the questions.

Element	% by mass of element in lithosphere	% by volume of element in atmosphere
Aluminium	8	0
argon	0	1
iron	5	0
nitrogen	0	78
oxygen	47	21
silicon	28	0

a Name the most abundant element in the lithosphere.

b Name the most abundant element in the atmosphere.

c Name the element that is abundant in both the atmosphere and the lithosphere.

d Name the most abundant non-metal element in the lithosphere.

e Name the most abundant metal in the lithosphere.

2 Draw lines to match each formula to a diagram.

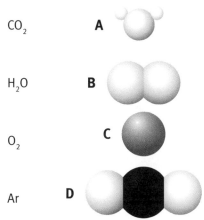

CO_2 **A**

H_2O **B**

O_2 **C**

Ar   **D**

3 Use words from the box to fill in the gaps. Each word may be used once, more than once, or not at all.

> **three   cannot   carbon   covalent**
> **silicon   four   slippery   hard**
> **can   lubricant   abrasive**

Diamond and graphite are two forms of the element _____. Diamond and graphite have giant _____ structures. In diamond, each atom is joined to _____ others by strong _____ bonds. These bonds mean that diamond is very _____. In graphite, each atom is joined to _____ others by strong _____ bonds to form layers. These layers can slide over each other, making graphite _____. This means it is a good _____. Graphite has electrons drifting between its layers, so it _____ conduct electricity.

4 Write word equations for the reactions below.

a zinc oxide reacting with carbon to make zinc and carbon dioxide

b copper oxide reacting with carbon

5 For each equation below, circle in **red** the reactant that is reduced. Circle in **blue** the reactant that is oxidised.

a iron oxide + carbon ⟶ iron + carbon dioxide

b tin oxide + carbon ⟶ tin + carbon dioxide

6 The stages below describe how aluminium is extracted from bauxite ore. They are in the wrong order.

A Add sodium hydroxide solution to remove impurities from the bauxite.

B Pass an electric current through the aluminium oxide.

C Melt the aluminium oxide.

D Aluminium forms at the negative electrode and oxygen forms at the positive electrode.

E Collect liquid aluminium from the tapping hole at the bottom of the tank.

Fill in the boxes to show the right order. The first one has been done for you.

A				

7 Draw lines to match each substance with its **type of structure** and its **melting point**. One has been done for you.

Substance	Type of structure	Boiling point (°C)
nitrogen	giant covalent	4830
silicon dioxide	giant ionic	−196
sodium chloride	giant covalent	2230
graphite	simple covalent	1413

H 8 Balance the equations below.

a $ZnO(s) + C(s) \longrightarrow Zn(l) + CO_2(g)$

b $Fe_2O_3(s) + C(s) \longrightarrow Fe(l) + CO_2(g)$

c $CuO(s) + C(s) \longrightarrow Cu(l) + CO_2(g)$

9 Write ionic equations, including state symbols, for:

a the reaction of lead nitrate solution with potassium iodide solution to make a precipitate of lead iodide, PbI_2

b the reaction of copper(II) chloride solution with sodium hydroxide solution to make a precipitate of copper hydroxide, $Cu(OH)_2$.

10 Calculate the mass of the metal in:

a 162 tonnes of zinc oxide, ZnO

b 51 kg of aluminium oxide, Al_2O_3.

1 Sarah needs to identify two white salts, A and B. She does the tests in the table opposite.

Test number	Test	Salt A observations	Salt B observations
1	Dissolve a little of the salt in water. Add a small volume of sodium hydroxide solution.	white precipitate	white precipitate
2	Add more sodium hydroxide solution.	no change	precipitate dissolves
3	Add hydrochloric acid to the solid.	no change	no change
4	Dissolve a little of the salt in water. Add dilute nitric acid, then barium nitrate solution.	no change	white precipitate
5	Add dilute nitric acid, then silver nitrate solution.	cream precipitate	no change

a Give the formula of the metal ion present in salt B.

_____ [1]

b Which of the two salts is a sulfate?

_____ [1]

c Give the name of salt A. Explain how you work out your answer.

_____

_____ [2]

> **Exam tip**
>
> You do not have to remember the results of the tests for ions – you will be given a data sheet to help you.

d Explain why nitric acid is added in test 5, and not hydrochloric acid.

_____

_____ [2]

H e Silver ions (Ag^+) react with chloride ions (Cl^-) to make silver chloride precipitate.

Write an ionic equation for this reaction.

Include state symbols.

_____ [2]

Total [8]

2 Many dental drills are made from diamond.

Diamond is a form of the element carbon.

Explain why the properties of diamond make it a suitable material for drill tips.

Use your knowledge and understanding of the structure and bonding of diamond to explain why diamond has these properties.

The quality of written communication will be assessed in your answer to this question.

Write your answer on separate paper or in your exercise book.

Total [6]

3 **a** The bar chart shows the melting points of four substances.

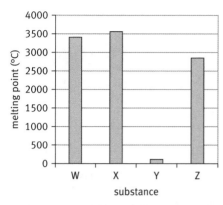

  i Substance Y exists as small molecules.
  Explain how the data on the bar chart show this.

  _____ [1]

  ii Which of the elements below could be substance Y?
  Use the periodic table to help you decide.
  Draw a ring around the correct answer.

  magnesium   iodine   technetium   antimony   [1]

b The table gives more data for substances W, X, and Z.

  i Which of the substances in the table might have a
  giant covalent structure?

  Give a reason for your decision.

  _____

  _____ [1]

Substance	Does it conduct electricity when solid?
W	yes
X	no
Z	no

  ii Suggest what further data you would need to help you
  decide which of the substances in the table is an ionic
  compound.

  Explain how this extra data would help you to make
  your decision.

  _____

  _____ [2]

c One of the substances in the table is tungsten.

  i Use the periodic table to help you identify which
  substance is tungsten. Draw a ring around the
  correct answer.

  substance W   substance X   substance Z   [1]

  ii Tungsten is used to make electrodes for electrolysis.
  Give one property of tungsten that makes it suitable
  for this purpose.

  _____ [1]

  Total [7]

4 **a** Magnesium bromide is an ionic compound.

  It used to be used as a sedative. It is also a laxative.

  i In which 'sphere' are many ionic compounds found
  dissolved in water?

  Tick one box.

  lithosphere   ☐
  atmosphere   ☐
  hydrosphere   ☐ [1]

ii Draw lines to match each property of magnesium bromide with a reason.

Property
When solid, it does not conduct electricity.
It has a high melting point.
When solid, it forms crystals.
When liquid, it conducts electricity.

Reason
There are very strong attractive forces between the positive and negative ions.
The ions are arranged in a regular pattern.
The charged particles cannot move.
The charged particles can move independently.

[3]

Ⓗ iii Magnesium bromide contains these ions:

Mg^{2+} and Br^-

What is the formula of magnesium bromide? _____ [2]

b In World War Two, magnesium was extracted from seawater to make bombs.

The mass of magnesium ions in 1 m³ of seawater is 1.3 kg.

Calculate the volume of seawater that contains 100 kg of magnesium ions.

Answer = _____ [2]

c i Today, magnesium metal is manufactured by the electrolysis of liquid magnesium chloride.

Which statements correctly describe what happens during this process?

Tick the correct two boxes.

Liquid magnesium metal forms at the positive electrode. ☐

An electric current decomposes the electrolyte. ☐

An electric current passes through liquid magnesium chloride. ☐

Chlorine gas is made at the negative electrode. ☐ [2]

ii Explain why magnesium cannot be produced by heating magnesium oxide with carbon.

_____ [1]

Total [11]

5 Serotonin is a hormone. It carries messages in the human body. Changes in the amount of serotonin in a person's brain may affect their mood. The chemical formula of serotonin is $C_{10}H_{12}N_2O$.

a Calculate the relative formula mass of serotonin.

Show clearly how you work out your answer.

Relative atomic masses: H = 1; C = 12; N = 14; O = 16

_____

_____ [2]

> **Exam tip**
>
> Write out calculations carefully before you do them. Then check your working, and your answer.

b The chemical dopamine also carries messages in the brain.

Its formula is $C_6H_{11}NO_2$.

The relative formula mass of dopamine is 153.

Calculate the mass of nitrogen in 153 g of dopamine.

Show clearly how you work out your answer.

_____

_____ [2]

Total [4]

6 The lithosphere contains large amounts of silicon dioxide.

a Finish the sentences about the structure of silicon dioxide.

Choose words from this list.

| simple   giant   weak   strong |

Silicon dioxide has a _____ covalent structure.

Its atoms are held together by _____ covalent bonds. [2]

b This is a list of properties of silicon dioxide.

A It has a high boiling point.

B Solid and liquid silicon dioxide do not conduct electricity.

C It is very hard.

D It is insoluble in water.

E It has a high melting point.

C 5

Which properties best explain the following facts about silicon dioxide?

Choose from the letters **A, B, C, D,** and **E**.

i   Silicon dioxide is used as an abrasive to make surfaces smooth.

_____ [1]

ii   Silicon dioxide is used to line furnaces that heat things to very high temperatures.

_____ [1]

Total [4]

Going for the highest grades

7   Aluminium is used to make overhead power lines.

a   **i**   Draw a labelled diagram in the space below to represent the bonding in aluminium metal.

[2]

ii   Use your diagram to explain why aluminium is a good conductor of electricity, and why it is malleable.

_____

_____

_____ [2]

b   Aluminium is extracted from bauxite.
Bauxite is mainly aluminium oxide, Al_2O_3.

i   Calculate the mass of aluminium that can be extracted from 1000 tonnes of aluminium oxide.

_____

_____

_____ [3]

ii   Describe how aluminium is extracted from aluminium oxide by electrolysis. In your answer, include a description of what happens at each electrode.

The quality of written communication will be assessed in your answer to this question.

Write your answer on separate paper, or in your exercise book.   [6]

Total [13]

1 Write each example from the box in an appropriate place on the diagram.

food additives e.g.

_____

fertilisers e.g.

_____

Chemical synthesis provides chemicals for...

pharmaceuticals e.g.

_____

plastics e.g.

_____

pigments e.g. titanium oxide

paracetamol
polythene
saccharin (a sweetener)
ammonium nitrate

2 The pie chart shows the percentage value of products made by the chemical industry in the UK.

Complete the sentences.

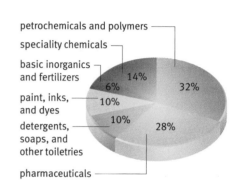

petrochemicals and polymers — 32%
speciality chemicals — 14%
basic inorganics and fertilizers — 6%
paint, inks, and dyes — 10%
detergents, soaps, and other toiletries — 10%
pharmaceuticals — 28%

a Chemicals used in _____ earn the most money for the British chemical industry.

b The total percentage value of paints, inks, dyes, and pharmaceuticals is _____.

c The percentage value of detergents, soaps, and other toiletries is _____% less than that of speciality chemicals.

3 For each reaction A, B, and C, decide which of the methods below you could use to measure the rate of reaction. Write the letters A, B, and C below the appropriate methods. You may choose more than one method for each reaction.

A $Mg(s) + H_2SO_4(aq) \longrightarrow MgSO_4(aq) + H_2(g)$

B $CaCO_3(s) + 2HNO_3(aq) \longrightarrow Ca(NO_3)_2(aq) + CO_2(g) + H_2O(l)$

C $Na_2S_2O_3(aq) + 2HCl(aq) \longrightarrow 2NaCl(aq) + SO_2(aq) + H_2O(l) + S(s)$

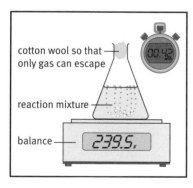

cotton wool so that only gas can escape

reaction mixture

balance — 239.5g

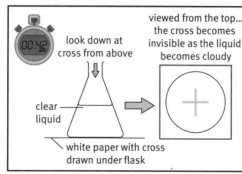

look down at cross from above

viewed from the top... the cross becomes invisible as the liquid becomes cloudy

clear liquid

white paper with cross drawn under flask

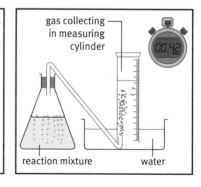

gas collecting in measuring cylinder

reaction mixture   water

C 6

4 The diagram shows apparatus for a titration.

Use the phrases in the box to label the diagram.

| accurately weighed | solid sample | pure water |
| titration flask | acid or alkali | burette |

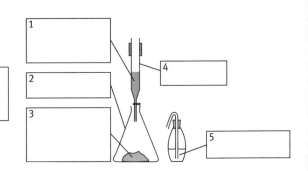

5 Solve the clues to fill in the grid.

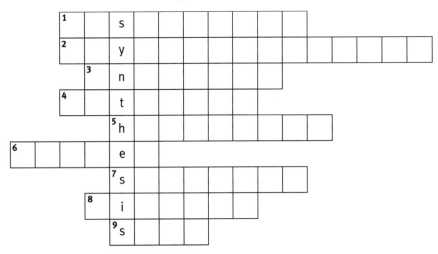

1 Use a _____ to finish drying a crystalline product of a synthesis.

2 Obtain a solid product from its solution by _____.

3 The point in a titration at which the reaction is just complete.

4 A chemical that speeds up a chemical reaction but is not used up in the process.

5 Alkaline solutions contain _____ ions.

6 Use more dilute solutions to make a reaction go _____.

7 React _____ acid with magnesium ribbon to make magnesium sulfate.

8 React nitric acid with calcium carbonate to make calcium _____.

9 A _____ is formed when an acid neutralises an alkali.

C6.1.1–2 Why is chemical synthesis important?

Chemical synthesis provides important chemicals for food additives, fertilisers, dyes, paints, and pharmaceuticals (medicines).

Bulk chemicals are made on a huge scale. They include ammonia, sodium hydroxide, sulfuric acid, and chlorine.

C6.1.7–10, C6.1.18–19 What are acids and alkalis?

An **acid** is a compound that dissolves in water to give a solution of pH less than 7. Acids produce hydrogen ions, $H^+(aq)$, in water.

Pure acidic compounds include:
* solids, for example, citric acid and tartaric acid
* liquids, for example, sulfuric acid, nitric acid, and ethanoic acid
* gases, for example, hydrogen chloride.

An **alkali** dissolves in water to give a solution of pH greater than 7. Alkalis produce hydroxide ions, $OH^-(aq)$, in water. Alkalis include potassium hydroxide, sodium hydroxide, and calcium hydroxide.

Indicators show whether a solution is acidic, alkaline, or neutral. **pH meters** also measure pH.

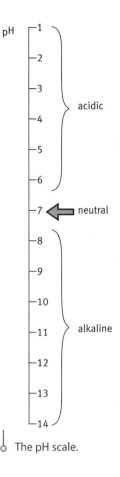
The pH scale.

C6.1.11, 16, 17, 20, 22 How do acids react?

Acids make **salts** in many of their reactions. Salts are ionic compounds that can be made from reactions of acids.
* **acid + metal ⟶ a salt + hydrogen**
  For example:
  hydrochloric acid + calcium ⟶ calcium chloride + hydrogen
  $$2HCl(aq) \quad + \quad Ca(s) \longrightarrow \quad CaCl_2(aq) \quad + \quad H_2(g)$$
* **acid + metal oxide ⟶ a salt + water**
  nitric acid + magnesium oxide ⟶ magnesium nitrate + water
  $$2HNO_3(aq) + \quad MgO(s) \quad \longrightarrow \quad Mg(NO_3)_2(aq) \quad + H_2O(l)$$
* **acid + metal carbonate ⟶ a salt + carbon dioxide + water**
  sulfuric acid + magnesium carbonate ⟶ magnesium sulfate + carbon dioxide + water
  $$H_2SO_4(aq) \quad + \quad MgCO_3(s) \quad \longrightarrow \quad MgSO_4(aq) \quad + \quad CO_2(g) \quad + H_2O(l)$$
* **acid + metal hydroxide ⟶ a salt + water**
  hydrochloric acid + sodium hydroxide ⟶ sodium chloride + water
  $$HCl(aq) \quad + \quad NaOH(aq) \quad \longrightarrow \quad NaCl(aq) \quad + H_2O(l)$$

  This is a **neutralisation** reaction. Hydrogen ions from the acid react with hydroxide ions from the alkali:
  $$H+(aq) + OH–(aq) \longrightarrow H_2O(l)$$

C
6

C6.4.21 How do we work out salt formulae?

Nitric acid (HNO_3) reacts with sodium hydroxide (NaOH) to make sodium nitrate. What is the formula of sodium nitrate?
- Charge on hydrogen ion = +1. So charge on nitrate ion = −1.
- Charge on hydroxide ion = −1. So charge on sodium ion = +1.

The total charge on the two ions in sodium nitrate is zero. So the formula has one Na^+ ion and one NO_3^- ion. The formula is $NaNO_3$.

C6.1.23–25 Why do energy changes matter?

Chemists need to know the energy changes in chemical reactions to minimise fuel costs and avoid accidents. Reactions can be:
- **exothermic** – they give out energy to the surroundings, making the temperature of the reaction mixture increase
- **endothermic** – they absorb energy from the surroundings, making the temperature of the reaction mixture decrease.

Energy-level diagrams show whether a reaction is exothermic or endothermic.

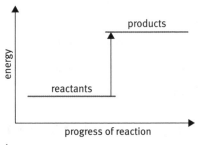

An exothermic reaction.

An endothermic reaction.

C6.2.5–6 What is relative atomic mass?

The **relative atomic mass** is the mass of an atom compared to the mass of a carbon atom. The relative atomic mass of carbon is 12. The relative atomic mass of helium is 4. So a carbon atom is 3 times heavier than a helium atom. The periodic table shows the relative atomic mass of every element.

C6.2.1a–f, C6.2.9–10 How can we do a chemical synthesis?

Choose the reaction to make the product

You need to make calcium chloride. You could make it by reacting hydrochloric acid with calcium *or* calcium oxide *or* calcium carbonate. Calcium carbonate is cheapest, so use the reaction:

calcium carbonate + hydrochloric acid $\longrightarrow$ calcium chloride + carbon dioxide + water

Do a risk assessment

Identify hazardous chemicals, and any other hazards, and take precautions to minimise risks from these hazards.

Work out the amounts of reactants to use

Add excess calcium carbonate to dilute hydrochloric acid.
- Calculate the relative formula masses of the reactants and calcium chloride.
- Calculate the relative reacting masses of these chemicals.

- Add units to convert to reacting masses.
  $CaCO_3(s) + 2HCl(aq) \longrightarrow CaCl_2(aq) + CO_2(g) + H_2O(l)$
  Relative formula masses:
  | 100 | 36.5 | 111 |
  Reacting masses:
  | 110 g | 36.5 g | 111 g |
- Scale reacting masses up or down to calculate amounts to use.
  To make 11.1 g of calcium chloride you need $(71 \div 10) = 7.1$ g of hydrogen chloride. This is the mass in 50 cm^3 of hydrochloric acid solution of concentration 142 g/litre.
  10 g of calcium carbonate reacts exactly with 7.1 g of hydrochloric acid. You need excess, so use about 12 g.

Do the reaction in suitable apparatus in the right conditions

- Use small lumps of the solid – powder reacts too fast.
- The excess calcium carbonate does not dissolve.

Separate the product from the reaction mixture

- Use **filtration** to separate the excess solid calcium carbonate from the product (calcium chloride solution).

Purify the product

- Heat gently in an **evaporating dish** to evaporate some water.
- Leave the concentrated solution to cool and crystallise.
- Put the crystals in an oven and then a **desiccator** to dry.

Measure the yield of the product

- The **actual yield** is the mass of product you made.
- The **theoretical yield** is the maximum possible yield of calcium chloride, calculated at the start.
- You can calculate the **percentage yield** like this:

$$\text{percentage yield} = \frac{\text{actual yield}}{\text{theoretical yield}} \times 100\%$$

C6.2.1g, 6.2.3, 6.2.11 How can we check the purity of a product?

Products must be of the correct purity for their purpose.

To check the purity of solid citric acid:
- Fill a burette with sodium hydroxide solution. Make sure you know its exact concentration.
- Accurately weigh a sample of the solid citric acid.
  Put it in a conical flask.

C 6

- Add pure water to the solid. Stir until it dissolves.
- Add a few drops of phenolphthalein indicator. This is colourless in acid solution.
- Add sodium hydroxide solution from the burette. Stop adding it when one drop of sodium hydroxide solution makes the indicator go pink.

C6.2.13–21 What are reaction rates?

The **rate of a reaction** is a measure of how quickly it happens. Chemists control reaction rates so that a reaction is not dangerously fast or uneconomically slow.

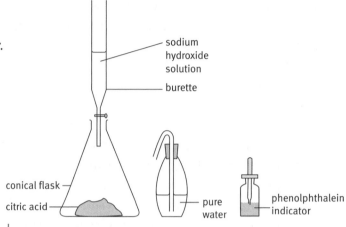

sodium hydroxide solution

burette

conical flask

citric acid

pure water

phenolphthalein indicator

⚬ Apparatus for measuring the purity of a sample of solid citric acid.

Measuring reaction rate

Methods for following reaction rate include:
- collecting a gas product and recording its volume regularly
- measuring mass decrease as a gas forms and escapes
- measuring the time for a known mass of solid to disappear
- measuring the time for a precipitate to hide a cross.

Factors affecting reaction rate

These factors affect the rate of a reaction:
- The **concentration** of reactants in solution – more concentrated solutions react faster.
- **Temperature** – increasing the temperature increases reaction rate.
- **Surface area** – 10 g of a powdered solid has a bigger surface area than 10 g of one lump of the same chemical. Increasing surface area increases the amount of contact between the solid and solution. This increases the reaction rate.
- A **catalyst** speeds up a reaction. It is not used up in the reaction.

lower concentration

higher concentration

⚬ Molecules have a greater rate of collision in a more concentrated solution.

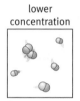

one big lump (slow reaction)

several small lumps (faster reaction)

C6.1.3–5, 6.1.12–15, 6.2.7 Fundamental chemistry

The Module C5 Fact bank shows how to calculate relative formula mass. Modules C4 and C5 give many opportunities to practise writing word equations and interpreting symbol equations.

H The Module C4 Factbank shows how to work out formulae of ionic compounds, and charges on ions. It also includes a section on balancing equations.

Formulae for many elements, compounds, and ions are also given.

Use extra paper to answer these questions if you need to.

1 Write the names of these pure acidic compounds in a bigger copy of the table.

tartaric, ethanoic, hydrogen chloride, nitric, citric, sulfuric

Gas	Liquid	Solid

2 For the sentences below:
- Write **acid** next to each sentence that is true for **acids**.
- Write **alkali** next to each sentence that is true for **alkalis**.
- Write **both** next to each sentence that is true for both **acids and alkalis**.

a They have a pH less than 7. _____
b They produce OH⁻ ions when they dissolve in water. _____
c They make litmus indicator turn red. _____
d They neutralise acids. _____
e Wear eye protection when working with these. _____
f Concentrated solutions of these are more dangerous than dilute solutions. _____
g Use a pH meter to measure the pH of solutions of these. _____

3 Katie added lumps of calcium carbonate to hydrochloric acid. The calcium carbonate and hydrochloric acid reacted together. Katie did the reaction 5 more times.
Tick one column in each row to show how the reaction rate changes each time.

Change	The reaction gets ...		
	faster	slower	can't tell
a Use bigger lumps of calcium carbonate.			
b Use more concentrated acid.			
c Heat the reaction mixture.			
d Use bigger lumps of calcium carbonate and heat the mixture.			
e Add a catalyst.			

4 Write definitions of the phrases below.
a relative atomic mass
b exothermic reaction

5 Riana dissolved 1.5 g of impure tartaric acid in pure water. She titrated with sodium hydroxide solution. At the end point, 19.0 cm³ of alkali had been added. Use the formula below to find the percentage purity of the product.

$$\% \text{ purity} = \frac{\text{titre} \times 7.58}{\text{mass of tartaric acid}}$$

6 Fill in the empty boxes. Use the periodic table to get the data you need.

Name of chemical	Formula	Relative formula mass
nitrogen gas		
nitric acid		
	$MgSO_4$	
	KCl	
calcium chloride		
	Na_2CO_3	
calcium carbonate		

7 Do calculations to fill in the empty boxes.

Formula of product	Actual yield	Theoretical yield	Percentage yield
SrO	98 kg	104 kg	
Al_2O_3	222 g	224 g	
SF_6	68 t	73 t	

H 8 a The formula of magnesium carbonate is $MgCO_3$. Carbonate ions have a charge of -2. What is the charge on a magnesium ion?
b The formula of aluminium oxide is Al_2O_3. The formula of an oxide ion is O^{2-}. What is the charge on an aluminium ion?

9 Fill in the empty boxes.

Name of salt	Formula of acid used to make the salt	Formula of hydroxide used to make the salt	Formula of salt
potassium chloride	HCl	KOH	
sodium sulfate	H_2SO_4	NaOH	
calcium nitrate	HNO_3	$Ca(OH)_2$	
lithium chloride	HCl	LiOH	

10 Balance the equations below. Then calculate the reacting masses for each of the substances shown in the equations.
a $NaOH + HCl \longrightarrow NaCl + H_2O$
b $KOH + H_2SO_4 \longrightarrow K_2SO_4 + H_2O$
c $Mg + O_2 \longrightarrow MgO$
d $Li + O_2 \longrightarrow Li_2O$
e $AgNO_3 + NaCl \longrightarrow AgCl + NaNO_3$
f $Pb(NO_3)_2 + KCl \longrightarrow PbCl_2 + KNO_3$
g $Fe_2O_3 + C \longrightarrow CO + Fe$
h $CaCO_3 \longrightarrow CaO + CO_2$

C 6

1   Mia reacts small lumps of calcium carbonate with dilute hydrochloric acid.

She uses the apparatus below.

a   i   Complete the equation below to show:

- the **formula** of the salt that is formed

- the **state symbol** for each of the chemicals at room temperature.

$$2HCl \underline{\quad} + CaCO_3(s) \longrightarrow \underline{\quad} (aq) + CO_2 \underline{\quad} + H_2O \underline{\quad}$$ [3]

ii   Mia makes 1.60 g of carbon dioxide.

Calculate the mass of calcium carbonate that reacted to make 1.60 g of carbon dioxide. Show your working.

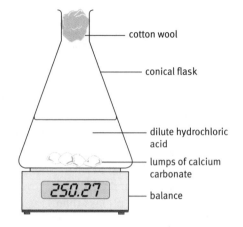

cotton wool

conical flask

dilute hydrochloric acid

lumps of calcium carbonate

balance

Mass of calcium carbonate = _____ g   [2]

b   Mia records the mass of carbon dioxide produced by the reaction every minute.

Her results are in the table below.

Time (minutes)	Mass of carbon dioxide produced since start (g)
0.0	0.00
1.0	1.10
2.0	1.40
3.0	1.56
4.0	1.60
5.0	1.60

Calculate the rate of reaction in the first minute. Show your working.

Rate of reaction in the first minute = _____ g/min
[1]

c   Mia does the experiment four more times.

Each time she changes one of the reaction conditions.

She calculates the rate of reaction during the first minute.

Her results are in the table below.

The rate that is missing from the table is the rate you calculated in answer to part b.

Experiment	Rate of reaction during first minute (g/min)
original experiment	
W	1.10
X	2.20
Y	0.70
Z	0.50

i   Give the letter of one experiment in which Mia might have used a less concentrated acid than in the original experiment.

Give a reason for your choice.

_____ [1]

ii   Suggest two changes to the conditions of the original experiment that Mia might have made in experiment X.

_____

_____ [2]

d   An energy-level diagram for the reaction of calcium carbonate with dilute hydrochloric acid is shown opposite.

What does the energy-level diagram tell you about the reaction?

_____

_____ [2]

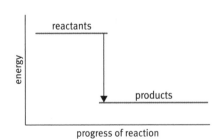

Total [11]

2   Two students investigated the claim on a carton of blackcurrant drink.

"Our blackcurrant drink has four times more vitamin C than orange juice."

The students first titrated 10.00 cm³ samples of **orange juice** with DCPIP solution. DCPIP reacts with vitamin C. The more DCPIP required, the greater the amount of vitamin C in the drink sample. The titration results of the two students are in the tables below.

Jude

	Run 1 (rough)	Run 2	Run 3	Run 4	Run 5
Initial burette reading (cm³)	1.00	13.60	25.50	37.50	2.50
Final burette reading (cm³)	13.60	25.50	37.50	49.60	17.50
Volume of DCPIP added (cm³)	12.60				

Lucy

	Run 1 (rough)	Run 2
Initial burette reading (cm³)	2.00	14.30
Final burette reading (cm³)	14.30	26.30
Volume of DCPIP added (cm³)	12.30	12.00

a Whose data are more likely to give a value for the volume of DCPIP required that is closest to the true value? Give a reason for your decision.

_____

_____ [1]

b i Calculate the volumes of DCPIP added for each run in Jude's titration. Write your answers in the table of Jude's results above.

ii Use your answer to part to identify the outlier in Jude's results.

_____ [1]

iii Jude decided to discard the outlier value. Suggest why.

_____ [1]

c Use the results in Jude's table to calculate the mean volume of DCPIP that reacts with 10.00 cm³ of orange juice. Include units in your answer.

_____

_____ [1]

d Jude then titrated 10.00 cm³ samples of the blackcurrant drink with DCPIP. The mean volume of DCPIP required was 1.50 cm³.

Do the results of Jude's titrations with orange juice and blackcurrant drink support the claim on the blackcurrant drink carton?

"Our blackcurrant drink has four times more vitamin C than orange juice."

Explain your answer.

_____

_____ [1]

Total [5]

3 a Raj wants to know the purity of a
sample of citric acid.
He does a titration to find out.
He uses this apparatus.

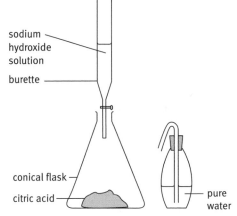

The stages below describe how Raj
does the titration.

They are in the wrong order.

A Accurately weigh out a sample
of citric acid. Put it in the conical
flask.

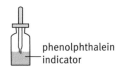

B Add sodium hydroxide solution from the burette, a few
cubic centimetres at a time. Swirl after each addition.

C Add sodium hydroxide solution from the burette, drop by
drop. Swirl after each addition.

D Add a few drops of phenolphthalein indicator. This is
colourless in acid solution.

E Stop adding sodium hydroxide solution when the indicator
is permanently pink.

F Add pure water. Stir until the solid dissolves.

Fill in the boxes to show the correct order.
The first one has been done for you.

| A | | | | | |

[3]

b The reaction in the titration is a neutralisation reaction.

i Complete the word equation for the reaction.

citric acid + sodium hydroxide $\longrightarrow$ sodium citrate + _____

[1]

ii The reaction can be represented by an **ionic equation**.
Complete the ionic equation.

H^+ + _____ $\longrightarrow$ _____

[2]

iii Give the name of the chemical that supplied the H^+ ions.

_____ [1]

Total [7]

4 Two students, Grace and Nzila, make copper sulfate crystals by reacting excess copper oxide powder with dilute sulfuric acid.

Below are the word and symbol equations for the reaction.

copper oxide + sulfuric acid ⟶ copper sulfate + water

$$CuO(s) \quad + \quad H_2SO_4(aq) \quad \longrightarrow \quad CuSO_4(aq) \quad + H_2O(l)$$

The students work separately.

Grace achieves a yield of 61% for her copper sulfate crystals. Nzila's yield is 92%.

Describe the steps for making the crystals. For each step, suggest a difference in technique between the two students that might explain their different yields.

✎ The quality of written communication will be assessed in your answer to this question.
Write your answer on separate paper or in your exercise book. **Total [6]**

> **Exam tip**
>
> If you're asked to write instructions for an experiment, jot down rough notes first to make sure you don't forget any of the stages.

⊕ Going for the highest grades

5 A company extracts mercury from its ore, mercury sulfide, by heating the ore in air.

This is the equation for the reaction.

$$HgS(s) + O_2(g) \longrightarrow SO_2(g) + Hg(l)$$

a Calculate the maximum mass of mercury that can be extracted from 233 kg of mercury sulfide.

Answer = _____ kg [2]

b In 2011, the company produced 1005 kg of mercury.

What mass of sulfur dioxide gas did the company produce as a by-product?

Answer = _____ kg [2]

c Sulfur dioxide can be used to make sulfuric acid.

This is a simplified summary of the process:

$$SO_2 + \tfrac{1}{2}O_2 \longrightarrow SO_3$$
$$SO_3 + H_2O \longrightarrow H_2SO_4$$

What is the maximum mass of sulfuric acid that the company could make from the sulfur dioxide produced in b?

Answer = _____ kg [2]

Total [6]

1 Finish the text in Dr Emerald's speech bubbles.

2 The equation below summarises the laboratory preparation of calcium chloride.

$CaCO_3(s) + 2HCl(aq) \longrightarrow CaCl_2(aq) + CO_2(g) + H_2O(l)$

Calculate:

the total number of atoms in the reactants	
the total relative atomic mass of these atoms	
the total number of atoms in the calcium chloride	
the total relative atomic mass of these atoms	
the atom economy for the laboratory preparation of calcium chloride made by this method	

3 Draw lines to link pairs of words or phrases.

You can draw more than one line from each word if you wish.

Write a sentence on each line to show how the two words or phrases are linked together.

bulk chemicals

sulfuric acid   fine chemicals

research and development   new processes

catalysts   government regulations

health and safety   medicinal drugs

food additives   storage

transport   ammonia

fragrances

C7.1.1–5 The chemical industry

Every year, the chemical industry makes **bulk chemicals** such as sulfuric acid, phosphoric acid, sodium hydroxide, and ammonia. Bulk chemicals are made continuously in very large quantities.

The industry makes **fine chemicals** such as medicinal drugs, food additives, and fragrances. These are made in small batches because fine chemicals are very high-quality products.

The chemical industry employs chemists for **research and development**. They aim to discover:
- new products, to meet a new need or to meet an existing need in a new way
- new processes, to make a product more cheaply or sustainably. This is one reason why chemists develop new catalysts.

Government health and safety regulations control chemical processes, and the storage and transport of chemicals. The regulations protect people (for example, workers at a factory) and the environment.

C7.1.6 Producing useful chemicals

The production of a chemical involves the following stages.
- The raw materials are converted into **feedstocks** (the starting materials for the process)
- The feedstocks are converted into products. This is **synthesis**. It happens in a reactor. Feedstocks may be fed into the reactor at high temperature or pressure. The reactor may contain a catalyst.
- A mixture of products, by-products, and waste may leave the reactor. The chemical plant **separates** these from each other. At this stage, unreacted feedstocks return to the reactor.
- The products are **analysed** to check their purity.

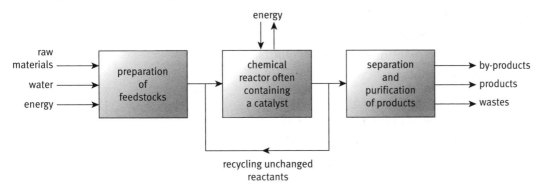

C7.1.7–10 Green chemistry

Green chemistry aims to make chemical processes as safe and sustainable as possible.

Chemists work towards achieving this aim by:

1 *using renewable feedstocks*, for example, making polyester from corn starch instead of crude oil.

2 *making the process as efficient as possible*; this is measured as the **atom economy**, which shows the mass of product atoms as a percentage of the mass of reactant atoms.

$$\text{atom economy} = \frac{\text{mass of atoms in the product}}{\text{mass of atoms in the reactants}} \times 100\%$$

For example, you can calculate the atom economy for the laboratory preparation of ethanoyl chloride like this:

$$3CH_3COOH + PCl_3 \longrightarrow 3CH_3COCl + H_3PO_3$$

Number of atoms in the reactants: 6C, 12H, 6O, 1P, 3Cl

(total relative atomic mass = 317.5)

Number of atoms ending up in the desired product: 6C, 9H, 3O, 3Cl

(total relative atomic mass = 235.5)

Number of atoms ending as waste: 3H, 1P, 3O

(total relative atomic mass = 82)

$$\text{atom economy} = \frac{235.5}{317.5} \times 100\%$$

$$= \textbf{74\%}$$

3 *reducing waste by*:
  • developing processes with higher atom economies
  • finding uses for by-products
  • increasing recycling at every stage of a product's life cycle.

4 *making processes more energy efficient by*:
  • insulating pipes and reaction vessels
  • using energy from exothermic reactions to heat reactants, generate electricity, or heat buildings.

5 *lowering the energy demand of a process by*, for example, developing processes that can run at lower temperatures. Every reaction has its own **activation energy**. This is the energy needed to break bonds in the reactants so that a reaction can start.

A catalyst provides an alternative route for a reaction, with a lower activation energy. The catalyst may reduce the energy inputs for the reaction and increase its rate.

> A process is sustainable if it uses resources in a way that can continue in future, replacing them as quickly as they are used. A process is renewable if the resources used can be replaced.

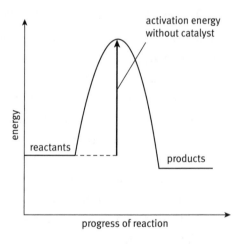

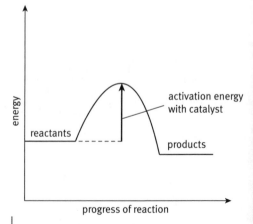

For a certain reaction, the activation energy is lower if a catalyst is added.

Some industrial processes are catalysed by natural catalysts, called **enzymes**. Every enzyme works best under particular conditions of temperature and pH. So if, for example, an enzyme for a reaction only works at temperatures below 37 °C, it is no good adding the enzyme to a reaction mixture at 300 °C.

6 *cutting pollution from wastes by*, for example:
- removing or destroying harmful chemicals before sending waste to the air, water, or landfill sites
- neutralising waste acids and alkalis
- precipitating toxic metal ions.

7 *avoiding hazardous chemicals by* replacing toxic feedstocks with chemicals that do not threaten human health or the environment.

8 *considering the social and economic impacts of a product and its manufacture*.

C7.1.11 Calculating formula masses

Example:

What is the formula mass of calcium carbonate?
- Write the formula: $CaCO_3$
- Use the periodic table to find the relative atomic masses of the elements in the compound: Ca = 40, C = 12, O = 12
- Add the relative atomic masses of all the atoms in the formula to find the relative formula mass: $40 + 12 + (16 \times 3) = 100$

C7.1.12 Calculating masses of reactants and products from balanced equations

Stage 1

Calculate the relative formula masses of the reactants and products.

Stage 2

Calculate the relative masses of reactants and products by taking into account the numbers used to balance the equation. Then add units to convert to reacting masses.

For example:

	Mg(s)	+	2HCl(aq)	$\longrightarrow$	$MgCl_2$(aq)	+	H_2(g)

Stage 1:

24	(1 + 35.5)	24 + (35.5 × 2)	1 × 2
= 24	= 36.5	= 95	= 2

Stage 2:

24 g	36.5 × 2 = **73 g**	**95 g**	**2 g**

Use extra paper to answer these questions if you need to.

1 For each pair of **bold** words below, highlight the word that is correct.

Bulk chemicals such as **ammonia / paracetamol** are made continuously in **large / small** amounts. Fine chemicals such as **food flavourings / sulfuric acid** are made in **large / small** batches. The quality of fine chemicals must be controlled **less / more** carefully than the quality of bulk chemicals. The exact specification of a fine chemical is **more / less** likely to change than that of a bulk chemical.

2 Colour the names of bulk chemicals red. Colour the names of fine chemicals blue.

ammonia	phosphoric acid
food additives	sodium hydroxide
fragrances	medicinal drugs

3 The table below lists some factors that determine the sustainability of a chemical process. Copy and complete the table to explain how each factor affects the sustainability of a chemical process.

Factor	How this factor affects the sustainability of a chemical process
Can the feedstocks be replaced as quickly as they are being used?	
What is the atom economy of the process?	
What waste products are produced?	
Does the process produce useful by-products?	
What are the environmental impacts of the process?	
What are the health and safety risks of the process?	
What are the social impacts of the chemical made by the process?	
What are the economic impacts of the process?	

4 Write **T** next to the statements below that are **true**. Write corrected versions of the statements that are **false**.

a The activation energy of a reaction is the energy needed to make bonds to start a reaction.

b A catalyst provides a different route for a chemical reaction.

c Chemists develop new catalysts to speed up reactions.

d The activation energy of a catalysed reaction is higher than the activation energy of the same reaction without a catalyst.

e Many enzyme catalysts are denatured above temperatures of about 25 °C.

f Enzyme catalysts work well whatever the pH.

5 Calculate the relative formula masses of the compounds below. Use the periodic table to obtain the relative atomic masses of the elements in the compounds.

a calcium carbonate, $CaCO_3$

b sulfuric acid, H_2SO_4

c potassium manganate, $KMnO_4$

d silver nitrate, $AgNO_3$

e copper sulfate, $CuSO_4$

f sulfur dioxide, SO_2

g barium nitrate, $Ba(NO_3)_2$

h magnesium nitrate, $Mg(NO_3)_2$

6 The stages below describe the large-scale production of a chemical. They are in the wrong order.

A analyse the products to check purity

B synthesise products from reactants

C separate products from by-products and waste

D make feedstocks from raw materials

Write a letter A–D in each box to show the correct order.

H 7 Work out the reacting masses of each reactant and product in the equations below.

a $CuCO_3 \longrightarrow CuO + CO_2$

b $NaOH + HCl \longrightarrow NaCl + H_2O$

c $2Mg + O_2 \longrightarrow 2MgO$

d $2NaOH + H_2SO_4 \longrightarrow Na_2SO_4 + 2H_2O$

e $2HCl + CaCO_3 \longrightarrow CaCl_2 + CO_2 + H_2O$

f $2HCl + Mg \longrightarrow MgCl_2 + H_2$

g $Pb(NO_3)_2 + 2KI \longrightarrow PbI_2 + 2KNO_3$

h $2KOH + H_2SO_4 \longrightarrow K_2SO_4 + 2H_2O$

8 Calculate the minimum masses of copper oxide and sulfuric acid needed to make 15.95 g of copper sulfate by the reaction shown in the equation below. Use data from the periodic table.

$CuO(s) + H_2SO_4(aq) \longrightarrow CuSO_4(aq) + H_2O(l)$

9 Look at this equation.

$CaCO_3(s) + 2HCl(aq) \longrightarrow CaCl_2(aq) + CO_2(g) + H_2O(l)$

Use the equation and data from the periodic table to calculate:

a the maximum mass of calcium chloride that could be made starting with 100 g of calcium carbonate

b the mass of waste carbon dioxide that would be made starting from 100 g of calcium carbonate.

10 For the reaction below, calculate the masses of silver nitrate and sodium iodide that react together to make 2.35 g of silver iodide.

$AgNO_3(aq) + NaI(aq) \longrightarrow AgI(s) + NaNO_3(aq)$

1 Orange groves could soon fuel Spanish cars as technology is developed to turn the peel from the fruit into ethanol.

Spain's Valencia region produces four million tonnes of oranges each year. Most of the oranges are made into juice. Juicing creates over 240,000 tonnes of waste, most of which is sold as animal feed.

A promising new technology produces 80 litres of ethanol from one tonne of orange waste. Valencia plans to produce and sell ethanol fuel locally, so creating 2500 jobs and cutting the cost of fuel to motorists by 40%. Petrol imports to the region will fall by at least 40%.

Scientists calculate that fuelling cars with ethanol from oranges — instead of petrol — may reduce greenhouse gas emissions by up to 90%. Ethanol produces less carbon monoxide than petrol, too.

The diagram below shows the production process.

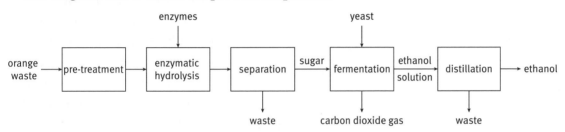

a The feedstock for the process is orange waste.

   i Is the feedstock renewable? Give a reason for your answer.

   _____

   _____ [1]

   ii Give one disadvantage of using orange waste as a feedstock for this process.

   _____

   _____ [1]

   iii Food crops such as maize can also be converted into ethanol for fuel. Suggest one advantage of using orange waste — and not food crops — as a feedstock for ethanol production.

   _____

   _____ [1]

b Give one social benefit of the Spanish project.

   _____ [1]

c **i** Give two environmental benefits of the Spanish project.

   _____

   _____ [2]

ii The ethanol production plant will be near the orange trees and the juice factory. Suggest how this will help to minimise the environmental impact.

_____

_____ [1]

iii One economic benefit of the project is cheaper fuel for motorists. Suggest how this could be a disadvantage for the environment.

_____

_____ [1]

d Of all the production stages, distillation needs the greatest energy input. Scientists are researching how to use waste from the separation stage to provide energy for the distillation stage. Suggest how this will make the overall process more sustainable.

_____

_____ [1]

e The process shown in the equation below converts glucose into ethanol by fermentation.

glucose $\xrightarrow{\text{YEAST}}$ ethanol + carbon dioxide

i Give one environmental problem caused by the by-product of this reaction.

_____ [1]

ii Complete and balance the symbol equation for the fermentation reaction.

$$C_6H_{12}O_6 \longrightarrow C_2H_5OH + \text{_____}$$ [2]

iii At what temperature is fermentation most efficient?

_____ [1]

iv Why must the fermentation products be distilled to make ethanol for fuel?

_____ [1]

Total [14]

2 The information in the box describes two methods of obtaining aluminium metal. Evaluate the sustainability of the two methods.

The quality of written communication will be assessed in your answer to this question.

Write your answer on separate paper or in your exercise book.

Total [6]

Recycling aluminium
A local council uses lorries to collect used aluminium cans from outside homes. A recycling company shreds the cans. It heats the shreds until they melt, and pours the liquid aluminium into a mould. The liquid cools and solidifies. Approximately 15 MJ of energy is needed to make 1 kg of aluminium by this method.

Extracting aluminium from its ore
Aluminium ore is crushed and processed to obtain pure aluminium oxide. At this stage large amounts of 'red mud' pollution are produced. The aluminium oxide is dissolved in a special solvent. A large electric current passes through the solution. The electricity splits up aluminium oxide to make liquid aluminium and other products. Approximately 260 MJ of energy is needed to make 1 kg of aluminium by this method.

1   Use words from the box to fill in the gaps on the pictures.

| ethanol   ethanoic acid   an ester   an alkane   methanol |

Beer contains _____

Pineapple-scented shampoo contains _____

Pickled onions preserved in _____

Butane camping gas contains _____

Car windscreen wash made from _____

2   The diagram below shows the structure of a molecule of a fat.

$$\begin{array}{l}
H \\
| \\
H-C-O-\overset{\overset{O}{\parallel}}{C}-CH_2-CH_2-CH_2-CH_2-CH_2-CH_2-CH_2-CH_2-CH_2-CH_2-CH_2-CH_3 \\
| \\
H-C-O-\overset{\overset{O}{\parallel}}{C}-CH_2-CH_2-CH_2-CH_2-CH_2-CH_2-CH_2-CH_2-CH_2-CH_2-CH_2-CH_3 \\
| \\
H-C-O-\overset{\overset{O}{\parallel}}{C}-CH_2-CH_2-CH_2-CH_2-CH_2-CH_2-CH_2-CH_2-CH_2-CH_2-CH_2-CH_3 \\
| \\
H
\end{array}$$

Draw a ⟨ring⟩ around each correct bold word.

An **animal** / **plant** made this fat to store **protein** / **energy**.
The fat is **saturated** / **unsaturated**. This means that **more** /
no more hydrogen atoms can be added to it.

Fats and oils are **alkanes** / **esters** of **butanol** / **glycerol** and
fatty / **amino** acids.

3 Make up 12 sentences using the phrases in the table.

Each sentence must include a phrase from each column.

Write your answers in the grid at the bottom.

For example, the sentence:

> Ethane ... doesn't react with aqueous solutions of
> acids and alkalis ... because ... C–C and C–H bonds are
> unreactive.

is

A	d	6

A Ethane	a reacts with sodium in a similar way to water		1 its –OH group interacts with water molecules.
B Ethanol	b does not react with sodium		2 it does not contain an –OH group.
C Octanol ($C_8H_{17}OH$)	c reacts with magnesium like other acids		3 it contains an –OH group.
D Ethanoic acid	d doesn't react with aqueous solutions of acids and alkalis		4 only a few of its molecules are ionised at any one time.
	e mixes easily with water	because	5 the attractions between hydrocarbon chains are weak.
	f does not mix with water		6 C–C and C–H bonds are difficult to break.
	g is a weak acid		7 it releases H^+ ions in solution.
	h has a lower boiling point than water		8 its hydrocarbon chain does not interact with –OH groups.

C7.2.1–9 Alkanes

The **alkanes** are a family of **hydrocarbons**. Hydrocarbons are made up of carbon and hydrogen only.

Name	Molecular formula	Structural formula
methane	CH_4	H | H—C—H | H
ethane	C_2H_6	H H | | H—C—C—H | | H H
propane	C_3H_8	H H H | | | H—C—C—C—H | | | H H H
butane	C_4H_{10}	H H H H | | | | H—C—C—C—C—H | | | | H H H H

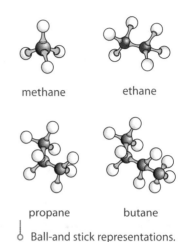

methane   ethane

propane   butane

Ball-and stick representations.

Alkanes burn in air. In a good supply of air, the combustion products are carbon dioxide and water. For example:

methane + oxygen $\longrightarrow$ carbon dioxide + water

$$CH_4(g) + 2O_2(g) \longrightarrow CO_2(g) + 2H_2O(g)$$

The symbol equation shows that one molecule of methane reacts with two molecules of oxygen to make one molecule of carbon dioxide and two molecules of water.

Alkanes do not react with solutions of acids or alkalis. This is because their C–C and C–H bonds are difficult to break and are therefore unreactive.

In alkanes, all the bonds between carbon atoms are single bonds (C–C). The molecules are **saturated** – no more hydrogen atoms can be added to them. The molecules of compounds in some other hydrocarbon families include double bonds between carbon atoms (C=C). These compounds are **unsaturated**.

C7.2.10–16 Alcohols

The alcohols are another family of organic compounds. Alcohol molecules contain an **–OH functional group**.

Name	Molecular formula	Structural formula	Uses
methanol	CH_3OH	H | H—C—O—H | H	raw material to make glues, foams, and windscreen wash
ethanol	C_2H_5OH	H H | | H—C—C—O—H | | H H	solvents, fuel, drinks

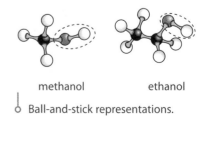

methanol   ethanol

Ball-and-stick representations.

Physical properties of alcohols

Methanol and ethanol mix well with water because their –OH group is similar to water, H_2O. Alcohols with long hydrocarbon chains, like heptanol ($C_7H_{15}OH$), do not mix with water. The long hydrocarbon chain does not interact with water molecules.

Ethanol boils at 79 °C. Propane, with a similar relative formula mass, boils at −42 °C. The alcohol has a higher boiling point because –OH groups tend to pull its molecules together.

Ethanol has a lower boiling point than water. There are only weak attractive forces between the hydrocarbon parts of ethanol molecules. So ethanol molecules are held together less strongly than water molecules.

Chemical properties of alcohols

Alcohol molecules have a hydrocarbon chain, so they burn in air:

ethanol   +   oxygen $\longrightarrow$ carbon dioxide   +   water

C_2H_5OH + $3O_2$ $\longrightarrow$ $2CO_2$ + $3H_2O$

Alcohols react with sodium. For example:

ethanol   +   sodium $\longrightarrow$ sodium ethoxide   +   hydrogen

sodium ethoxide

Water reacts with sodium in a similar way.

water + sodium $\longrightarrow$ sodium hydroxide + hydrogen

The reactions are similar because both water and ethanol molecules have an –OH group. When they react, an O–H bond breaks. Sodium ethoxide and sodium hydroxide are salts.

Alkanes do not react with sodium. This is because alkanes have no –OH groups, and their C–C and C–H bonds are unreactive.

C7.2.17–22 Making ethanol

The chemical industry makes large amounts of ethanol. Most of it is used for fuel. There are three main methods of manufacturing ethanol.

(1) From sugars

Fermentation converts simple sugars, such as glucose, into ethanol.

glucose $\longrightarrow$ ethanol   +   carbon dioxide

$C_6H_{12}O_6$ $\longrightarrow$ $2C_2H_5OH$ + $2CO_2$

The sugars come from plants such as sugar cane, sugar beet, maize, and rice.

Enzymes in yeast catalyse fermentation. The process works best between 25 °C and 37 °C. Below this temperature, the reaction is too slow. Above this temperature, the enzymes denature. The pH for the reaction must be correct, since changes in pH change the shape of enzymes and make them less effective.

The maximum concentration of an ethanol solution made by fermentation is about 14%. Above this concentration, yeast enzymes do not work.

An ethanol solution can be concentrated by **distillation** to make products like brandy and whisky.

(2) From waste biomass

Genetically modified *E. coli* bacteria can be used to convert waste biomass, for example, corn stalks and wood waste, into ethanol. The optimum conditions for the process are:
* temperature between 25 °C and 37 °C
* pH between 6 and 7.

(3) From ethane

Ethane is obtained from natural gas. It is converted into ethene in cracking reactions. Ethene is also obtained by cracking naphtha from crude oil.

Then ethene reacts with steam at 300 °C and 60–70 times atmospheric pressure. A phosphoric acid catalyst speeds up the reaction.

$$C_2H_4 + H_2O \longrightarrow C_2H_5OH$$

The first time ethene and steam go through the reactor, the yield is only 5%. So unreacted ethene and steam are sent through the reactor several times until the final yield is around 95%.

There are advantages and disadvantages to the three methods of producing ethanol. They are listed in the table below.

Method	Advantages	Disadvantages
fermentation of glucose	• feedstock renewable • low energy costs	• produces waste carbon dioxide • land used to grow crops for making ethanol, not food
conversion of waste biomass by genetically modified *E. coli*	• feedstock renewable • can use waste biomass as feedstock • low energy costs	• produces waste carbon dioxide • not yet used on a large scale
reaction of ethene with steam	• feedstock a by-product of cracking reaction of naphtha	• feedstock not renewable • high energy costs

C7.2.23–30 Carboxylic acids

Carboxylic acids contain this functional group:

Two examples of carboxylic acids are methanoic acid, molecular formula HCOOH, and ethanoic acid, CH_3COOH.

Vinegar is a dilute solution of ethanoic acid in water. Some carboxylic acids smell and taste bad – including those in sweaty socks and rancid butter.

The properties of carboxylic acids are determined by their functional group. Hydrogen ions leave this group when a carboxylic acid dissolves in water.

methanoic acid   ethanoic acid

Carboxylic acids.

ethanoic acid   ethanoate ion

Like other acids, carboxylic acids react with metals, alkalis, and carbonates:

ethanoic acid + sodium $\longrightarrow$ sodium ethanoate + hydrogen

methanoic acid + sodium hydroxide $\longrightarrow$ sodium methanoate + water

ethanoic acid + calcium carbonate $\longrightarrow$ calcium ethanoate + carbon dioxide + water

Carboxylic acids are **weak acids** – only a few molecules are ionised at any one time. This means they are less reactive than **strong acids** such as hydrochloric acid, sulfuric acid, and nitric acid.

Dilute solutions of weak acids have higher pH values than solutions of strong acids of the same concentration.

C7.2.31–36 Esters

Esters contain this functional group:

Esters have distinctive smells. They are responsible for the smells and flavours of fruits.

We use esters:
* to flavour foods
* to make sweet-smelling perfumes, shampoos, and bubble baths
* as solvents
* as plasticisers to make polymers such as PVC soft and flexible.

You can make an ester by warming a mixture of an alcohol and a carboxylic acid. You also need a catalyst, for example, sulfuric acid.

This equation summarises the reaction for making methyl ethanoate:

ethanoic acid + methanol $\longrightarrow$ methyl ethanoate + water

H Making an ester involves:
- heating the reactants under reflux
- distillation, to separate the ester from the reaction mixture
- purification, by shaking with reagents in a tap funnel
- drying, by shaking with anhydrous calcium chloride.

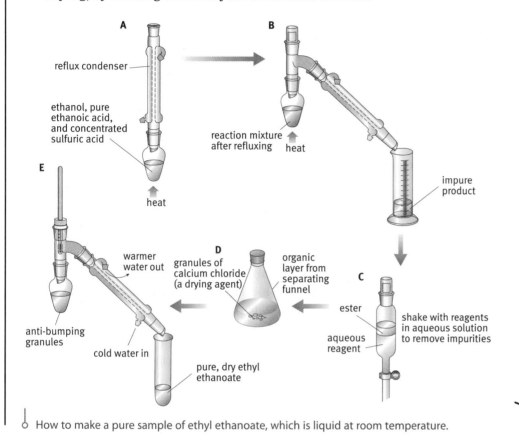

How to make a pure sample of ethyl ethanoate, which is liquid at room temperature.

C7.2.37–39 Fats and oils

Plants and animals make fats and oils as energy stores. Fats and oils are esters of:
- glycerol, an alcohol with three –OH groups
- fatty acids, which have a long hydrocarbon chain attached to a –COOH group.

The structure of a typical fat.

Animal fats, like butter and lard, are made up mostly of saturated molecules. They are usually solid at room temperature.

Vegetable oils, like sunflower oil, are made up mostly of unsaturated molecules. They are usually liquid at room temperature.

Use extra paper to answer these questions if you need to.

1 Draw lines to match each compound to one use.

Compound		Use
ethanol		to make glue
methanol		to make vinegar
ethanoic acid		as a fuel
ethane		as a food flavouring
pentyl pentanoate		to make ethanol

2 Identify one unpleasant smell and one unpleasant taste caused by carboxylic acids.

3 Draw lines to match the name of each compound to the family it belongs to.

Name of compound		Family
ethanoic acid		alkanes
ethanol		carboxylic acids
ethane		esters
ethyl ethanoate		alcohols

4 Write **T** next to the statements that are **true**. Write corrected versions of the statements that are **false**.

a Ethane burns in plenty of air to make ethanol and water.

b Alkanes do not react with acids because they contain only C–H and C–C bonds, which are easy to break and therefore unreactive.

c Ethanol mixes with water because its O–H group interacts with the O–H bonds in water.

d The limit to the concentration of ethanol solution that can be made by fermentation is about 94%.

e Ethanol solution can be concentrated by distillation to make whisky.

f *E. coli* converts waste biomass to ethanoic acid.

g A dilute solution of a weak acid has a lower pH than a solution of a strong acid of the same concentration.

h Carboxylic acids are stronger acids than hydrochloric acid.

5 Copy the table, filling in the empty boxes.

Name	Molecular formula	Structural formula
methane		
	C_4H_{10}	
propane		
	CH_3OH	

methanoic acid		$H-C\overset{O}{\underset{O-H}{}}$
	CH_3COOH	

6 Complete the word equations below.

a Propane + oxygen ⟶ _____ + water

b Ethanol + _____ ⟶ carbon dioxide + water

c Ethanoic acid + magnesium ⟶ magnesium ethanoate + _____

d Methanoic acid + _____ ⟶ magnesium methanoate + water

e Ethanoic acid + calcium carbonate ⟶ _____ + _____ + water

f Ethanol + ____ ⟶ sodium ethoxide + hydrogen

7 In each list below, circle two chemicals that react together to make an ester.

a Ethanoic acid, ethane, ethyl ethanoate, ethanol

b Propane, propanoic acid, methanol, methane

c Ethanol, butane, ethyl propanoate, propanoic acid

d Methane, methanol, methyl methanoate, methanoic acid

8 List some advantages and disadvantages of producing ethanol fuel by the fermentation of sugar cane compared with producing ethanol fuel from ethane.

H 9 Describe how ethanol reacts with sodium. Then compare this reaction to the reaction of ethane with sodium, and the reaction of water with sodium.

10 The stages below describe how to make pure, dry ethyl ethanoate in the laboratory.
They are in the wrong order.

A Put measured amounts of ethanol, ethanoic acid, and sulfuric acid in a flask.

B Shake with dilute sodium hydroxide solution to remove some impurities.

C Use distillation to separate ethyl ethanoate from the mixture.

D Pour the ethyl ethanoate into a separating funnel.

E Heat under reflux.

F Run the ethyl ethanoate into a flask.

G Add granules of anhydrous calcium chloride to dry the ethyl ethanoate.

Fill in the boxes to show the correct order. The first one has been done for you.

A						

11 Complete and balance the symbol equations for the reactions below.

a $CH_4 + O_2 \longrightarrow$

b $CH_3CH_2OH + Na \longrightarrow$

c $CH_3OH + O_2 \longrightarrow$

d $C_2H_6 + O_2 \longrightarrow$

1 Some crisp manufacturers make crisps by frying slices of potato in sunflower oil.

a Sunflower oil is extracted from sunflower seeds.

Why do plants make oils?

_____ [1]

b Sunflower oil is a mixture of oils. The oils are esters of glycerol and fatty acids.

The table gives information about some of these fatty acids.

Name of fatty acid	Is the fatty acid saturated or unsaturated?	Is the oil suitable for frying potatoes?
palmitic	saturated	yes
linoleic	unsaturated	no
oleic	unsaturated	yes
stearic	saturated	yes

Most scientists agree that eating too much saturated fat increases the risk of getting heart disease.

i Which of the structures below shows an oil that is an ester of glycerol and linoleic acid? _____

A

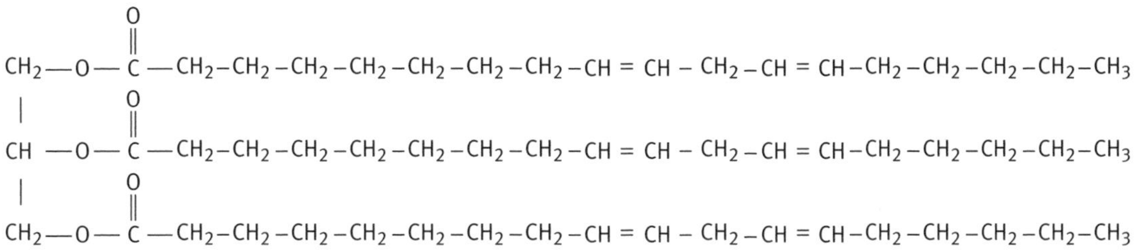

$$CH_2-O-\overset{\overset{O}{\|}}{C}-CH_2-CH_2-CH_2-CH_2-CH_2-CH_2-CH_2-CH=CH-CH_2-CH=CH-CH_2-CH_2-CH_2-CH_2-CH_3$$
$$CH-O-\overset{\overset{O}{\|}}{C}-CH_2-CH_2-CH_2-CH_2-CH_2-CH_2-CH_2-CH=CH-CH_2-CH=CH-CH_2-CH_2-CH_2-CH_2-CH_3$$
$$CH_2-O-\overset{\overset{O}{\|}}{C}-CH_2-CH_2-CH_2-CH_2-CH_2-CH_2-CH_2-CH=CH-CH_2-CH=CH-CH_2-CH_2-CH_2-CH_2-CH_3$$

B

$$CH_2-O-\overset{\overset{O}{\|}}{C}-CH_2-CH_2-CH_2-CH_2-CH_2-CH_2-CH_2-CH_2-CH_2-CH_2-CH_2-CH_2-CH_2-CH_2-CH_2-CH_2-CH_3$$
$$CH-O-\overset{\overset{O}{\|}}{C}-CH_2-CH_2-CH_2-CH_2-CH_2-CH_2-CH_2-CH_2-CH_2-CH_2-CH_2-CH_2-CH_2-CH_2-CH_2-CH_2-CH_3$$
$$CH_2-O-\overset{\overset{O}{\|}}{C}-CH_2-CH_2-CH_2-CH_2-CH_2-CH_2-CH_2-CH_2-CH_2-CH_2-CH_2-CH_2-CH_2-CH_2-CH_2-CH_2-CH_3$$

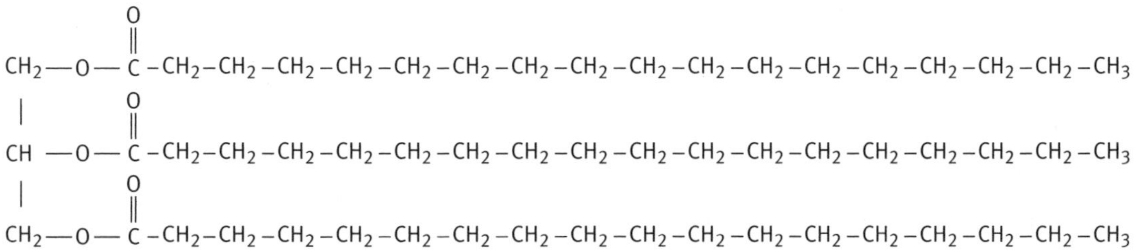

Give a reason for your choice.

_____ [1]

ii Until 1999, nearly all sunflower oil produced in the USA contained a high proportion of esters of linoleic acid.

Now, many US farmers grow sunflowers whose seeds produce oils containing a high proportion of esters of oleic acid.

Use the information in the box to suggest why.

_____

_____ [2]

Total [4]

2 The solvent in this nail varnish is ethyl ethanoate.

Ethyl ethanoate is an ester.

Nail-appeal varnish

a To make ethyl ethanoate, scientists react together two chemicals.

 i One of these chemicals is ethanol.

 Ethanol is an alcohol.

 Draw a (ring) around the formula of ethanol.

 C_2H_6 $\qquad$ C_2H_4 $\qquad$ C_2H_5OH $\qquad$ CH_3COOH $\qquad$ [1]

 ii The chemical that reacts with ethanol to make ethyl ethanoate is in another chemical family.

 Tick the box next to the name of this family.

 alcohols $\quad$ ☐ $\qquad$ alkanes $\quad$ ☐

 carboxylic acids $\quad$ ☐ $\qquad$ esters $\quad$ ☐ $\qquad$ [1]

 iii The scientists also add concentrated sulfuric acid to the reaction mixture.

 Explain why.

 _____ [1]

b Ethyl ethanoate occurs naturally in apples and bananas.

The ethyl ethanoate used to make nail varnish is not extracted from these fruits.

Suggest why.

_____

_____ [2]

Total [5]

⊕ Going for the highest grades

3 Two students, Esther and Jamie, make pure, dry ethyl ethanoate from ethanol and ethanoic acid. Below is the word equation for the reaction:

ethanol + ethanoic acid $\longrightarrow$ ethyl ethanoate + water

The students work separately. Esther achieves a yield of 60% for her ethyl ethanoate. Jamie's yield is 44%.

Describe the steps for making the ethyl ethanoate. For each step, suggest a difference in technique between the two students that might explain their different yields.

✎ The quality of written communication will be assessed in your answer to this question.

Write your answer on separate paper or in your exercise book.

Total [6]

1 Write the words and phrases in the correct places on the
   diagrams.

| exothermic | endothermic | photosynthesis | respiration |
| burning | less than | more than | lost to | gained from |

An _____ reaction   An _____ reaction

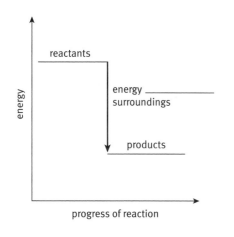

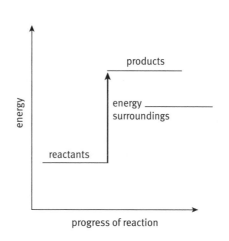

The total energy of the products is _____
the total energy of the reactants.

The total energy of the products is _____
the total energy of the reactants.

Examples: _____   Example: _____
           _____

2 Use the energy-level diagrams to answer the questions below.
   The scale is the same for each diagram.

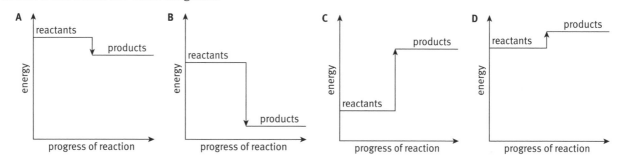

a Which diagrams represent exothermic reactions?

   _____

b Which diagrams represent reactions in which energy is
   given out? _____

c Which diagram represents the reaction that takes in
   most energy? _____

d Which diagram represents the reaction that gives out
   most energy? _____

e Which diagrams represent reactions in which the
   products have more energy than the reactants?

   _____

C7.3.1–2 Exothermic and endothermic reactions

Every chemical reaction involves energy changes.

Exothermic reactions give out energy to their surroundings. The energy of the products is less than the energy of the reactants.

Examples of exothermic reactions are:
- burning
- respiration
- the reaction of dilute hydrochloric acid with sodium hydroxide solution.

Endothermic reactions take in energy from their surroundings. The energy of the products is greater than the energy of the reactants.

Examples of endothermic reactions are:
- photosynthesis
- the reaction of solid citric acid with sodium hydrogencarbonate solution.

C7.3.3–4 Breaking and making bonds

In chemical reactions, chemical bonds in the reactants break. This process takes in energy. Bond breaking is endothermic.

Then new bonds are made to form products. This process gives out energy. Bond making is exothermic.

For example, when methane burns in oxygen:

$$
\begin{array}{c}
\underset{\overset{|}{H}}{\overset{\overset{H}{|}}{H-C-H}} \;+\; \begin{array}{c} O=O \\ O=O \end{array} \;\longrightarrow\; O=C=O \;+\; \begin{array}{c} {}^{H}\!\diagdown_{\,O} \\ {}_{H}\diagup \\ {}^{H}\!\diagdown_{\,O} \\ {}_{H}\diagup \end{array}
\end{array}
$$

In this example, the energy taken in to break bonds (in methane and oxygen) is less than the energy given out in making new bonds (in carbon dioxide and water). So the reaction is exothermic.

In endothermic reactions, the energy taken in to break bonds is more than the energy given out in making new bonds.

Ⓗ If you know the energy changes involved in breaking and making the bonds in a chemical reaction, you can estimate the overall energy change of the reaction.

For example, hydrogen reacts with fluorine to make hydrogen fluoride.

$H_2(g) + F_2(g) \longrightarrow 2HF(g)$

The energy for breaking and making the bonds in this reaction is shown below:

H–H	+	F–F	$\longrightarrow$	2H–F
434		158		$(2 \times 565) = 1130$ kJ

The energy taken in to break bonds in the reactants is
$434 + 158 = 592$ kJ

The energy given out in making bonds in the product is 1130 kJ

Overall energy = energy taken in − energy given out
change   to break bonds   in making bonds
− 538 kJ   = 592   − 1130 kJ
Overall energy change = − 538 kJ

The value for the energy transfer is negative. This shows that energy is given out to the surroundings. So the reaction of hydrogen with fluorine is exothermic.

C7.3.5 Activation energy

In a mixture of methane and oxygen, molecules collide all the time. But colliding molecules only react if they have enough energy to start breaking bonds.

Every reaction needs a certain minimum energy before it can start. This is the **activation energy**.

The diagrams represent the activation energies for two reactions of gases. Reaction A has a lower activation energy than reaction B.

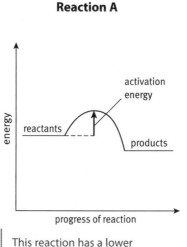

Reaction A

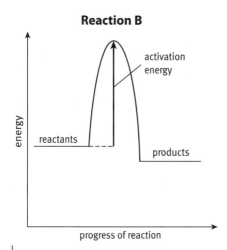

Reaction B

This reaction has a lower activation energy. The reaction is faster than reaction B at room temperature.

This reaction has a higher activation energy. The reaction is slower at room temperature.

Use extra paper to answer these questions if you need to.

1 Use words from the box to complete the sentences below. Each word may be used once, more than once, or not at all.

more	exothermic	endothermic	less

In an _____ reaction, energy is given out to the surroundings. The energy of the products is _____ than the energy of the reactants. In an _____ reaction, energy is taken in from the surroundings. The energy of the products is _____ than the energy of the reactants.

2 Draw straight lines to match each type of reaction to two examples.

Type of reaction	Example
exothermic	photosynthesis
	combustion
endothermic	respiration
	the reaction of citric acid with sodium hydrogencarbonate

3 For the sentences below:
- write **exo** next to each sentence that is true for **exothermic** reactions
- write **endo** next to each sentence that is true for **endothermic** reactions
- write **both** next to each sentence that is true for both **exothermic and endothermic** reactions.

a The energy of the products is less than the energy of the reactants.

b During the reaction, energy is transferred to the surroundings.

c The energy of the reactants is less than the energy of the products.

d During the reaction, energy is gained from the surroundings.

e Energy is needed to break bonds in the reactants.

f Energy is given out when new bonds are made to form products.

g During the reaction, energy is given out.

h The energy of the products is more than the energy of the reactants.

4 Write a **T** next to each statement that is **true**. Write corrected versions of the statements that are **false**.

a When molecules collide, they always react.

b Every reaction needs a certain minimum energy before it can start.

c The activation energy is the energy needed to break bonds to start a reaction.

d The activation energy is the same for every reaction.

5 Look at the energy-level diagram below.

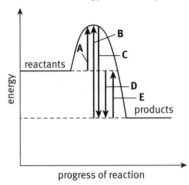

a Give the letter of the arrow that shows the activation energy for the reaction.

b Give the letter of the arrow that shows the overall energy change for the reaction.

c Is the reaction exothermic or endothermic? Explain how you decided.

6 For each reaction below, draw all the bonds that break and all the bonds that are made in the correct boxes.

a $H_2(g) + Cl_2(g) \longrightarrow 2HCl(g)$

bonds that break	bonds that are made

b $CH_4(g) + Cl_2(g) \longrightarrow CH_3Cl(g) + HCl(g)$

bonds that break	bonds that are made

H7 The table shows the energy needed to break some covalent bonds. Use the data in the table to state whether each of the statements below is **true** or **false**.

Bond	Energy needed to break bond for one relative formula mass (kJ)
H–H	434
O=O	497
H–O	463
Cl–Cl	243
H–Cl	431
C–H	413
C–Cl	339
O–O	146

a A double bond is easier to break than a single bond.

b The H–O bond is harder to break than the H–H bond.

c The H–Cl bond is stronger than the H–O bond.

d More energy is needed to break the C–H bond than the C–Cl bond.

8 Calculate energy changes for the reactions below. Use your answer to question 6, and the data in question 7.

a $H_2(g) + Cl_2(g) \longrightarrow 2HCl(g)$

b $CH_4(g) + Cl_2(g) \longrightarrow CH_3Cl(g) + HCl(g)$

c $2H_2(g) + O_2(g) \longrightarrow 2H_2O(g)$

C 7

1 Hydrogen iodide decomposes to make hydrogen and iodine.

Here is an equation for the reaction:

$$2HI \longrightarrow H_2 + I_2$$

The activation energy of the reaction is 183 kJ for the amounts shown in the equation.

a Explain the meaning of the term *activation energy*.

_____

_____ [1]

b Use a ruler to draw an *arrow* on the energy-level diagram below to show the activation energy of the reaction. Label the arrow 'activation energy'.

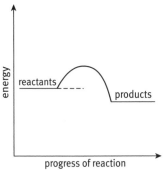

[1]

Total [2]

2 The chemical industry produces hydrogen by reaction 1 below.

Reaction 1: $CH_4(g) + H_2O(g) \longrightarrow CO(g) + 3H_2(g)$

The industry also produces hydrogen by the electrolysis of water (reaction 2).

The energy change for reaction 2 is +286 kJ for the amounts shown in the equation.

Going for the highest grades

Reaction 2: $H_2O(l) \longrightarrow H_2(g) + ½ O_2(g)$

Use the data in the table, and the data and equations above, to suggest and explain one reason why the chemical industry produces more hydrogen by reaction 1 than by reaction 2.

The quality of written communication will be assessed in your answer to this question.

Write your answer on separate paper or in your exercise book.

Bond	Average bond energy for one relative formula mass (kJ)
C–H	413
O–H	463
C≡O	1077
H–H	434

Total [6]

Ⓗ Going for the highest grades

3   Hydrogen and chlorine react to make hydrogen chloride.

Here is an equation for the reaction:

$$H — H \; + \; Cl — Cl \longrightarrow \begin{array}{c} H — Cl \\ H — Cl \end{array}$$

The table shows the average bond energies of the bonds in the reactants and product.

Bond	Average bond energy for one relative formula mass (kJ)
H–H	434
Cl–Cl	243
H–Cl	431

a   Use the values in the table to show that the energy change from breaking bonds is 677 kJ.

_____

_____ [1]

b   Calculate the energy change from making bonds for the amounts shown in the equation.

Use the equation for the reaction and data from the table.

_____

_____ [2]

c   i   For any reaction:

Energy transfer
= (energy change for breaking bonds) − (energy change for making bonds)

Work out the energy transfer for the reaction.

_____ [2]

ii   Explain how your value shows that the reaction is exothermic.

_____ [1]

iii   Complete the energy-level diagram on the right for the reaction of hydrogen with chlorine. You do not need to show the activation energy.   [2]

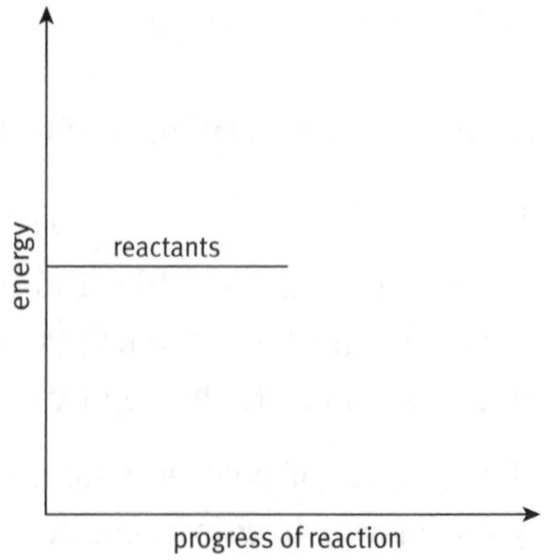

d   A scientist does an experiment to measure the energy transfer for the reaction. The size of the measured value is 1.5 kJ greater than the value calculated from the bond energies in the table.

Suggest why the measured and calculated values are not exactly the same.

_____ [1]

Total [9]

C
7

1 Fill in the tables for the equilibrium reactions below.

a $PCl_5 \rightleftharpoons PCl_3 + Cl_2$

Equation for forward reaction	
Equation for backward reaction	
Formulae of chemicals present at equilibrium	

b Write and balance a symbol equation for the word equation below:

nitrogen + hydrogen $\rightleftharpoons$ ammonia

Then complete the table below.

Equation for forward reaction	
Equation for backward reaction	
Formulae of chemicals present at equilibrium	

2 The diagram shows how ammonia is made by the Haber process.

Complete the labels by filling in the gaps.

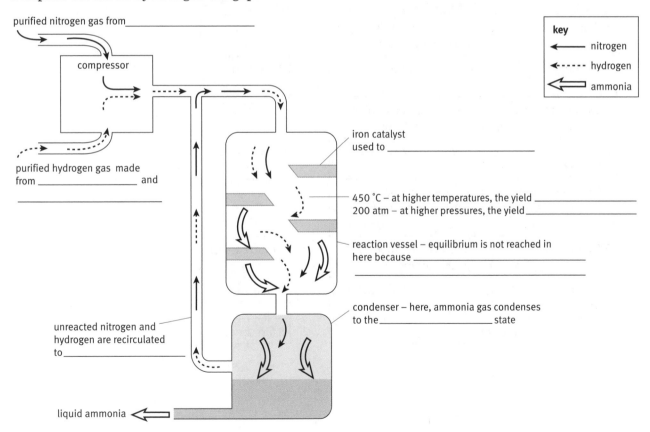

purified nitrogen gas from _____

compressor

key
→ nitrogen
←----- hydrogen
⇐ ammonia

iron catalyst used to _____

purified hydrogen gas made from _____ and
_____

450 °C – at higher temperatures, the yield _____
200 atm – at higher pressures, the yield _____

reaction vessel – equilibrium is not reached in here because _____
_____

condenser – here, ammonia gas condenses to the _____ state

unreacted nitrogen and hydrogen are recirculated to _____

liquid ammonia ⇐

3 Complete the labels by filling the gaps.

Use the words **higher** and **lower** only.

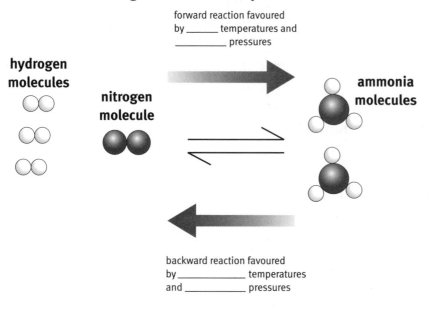

forward reaction favoured by _____ temperatures and _____ pressures

hydrogen molecules

nitrogen molecule

ammonia molecules

backward reaction favoured by _____ temperatures and _____ pressures

4 Solve the anagrams.

Match each anagram answer to a clue.

Clues	Anagrams
A The process of nitrogen _____ converts nitrogen gas from the air into nitrogen compounds.	1 ants tire
B Ammonia is a _____ made up of hydrogen and nitrogen.	2 taxi info
C Bacteria that live in the roots of a group of plants called _____ can fix nitrogen from the air.	3 a last cyst
D High levels of _____ in rivers and lakes make algae grow rapidly.	4 top blinder
E Some living organisms fix nitrogen at room temperature and pressure using enzymes as _____	5 greeny
F A nitrogen molecule is stable because there is a _____ _____ between its two atoms.	6 mod con up
G Chemists are trying to make new catalysts to fix nitrogen so as to reduce _____ costs.	7 slug mee
H Most ammonia is used to make _____ .	8 flirt series

C7.4.1 Reversible reactions

Many chemical reactions are **reversible**. They can go forwards or backwards. The direction of a reaction depends on conditions such as temperature, pressure, or the concentration of chemicals.

For example, at room temperature ammonia gas and hydrogen chloride gas react to make solid white ammonium chloride.

white smoke of ammonium chloride

ammonia	+	hydrogen chloride	$\longrightarrow$	ammonium chloride
$NH_3(g)$	+	$HCl(g)$	$\longrightarrow$	$NH_4Cl(s)$

If you gently heat solid ammonium chloride, it decomposes to make ammonia and hydrogen chloride gas again.

ammonium chloride	$\longrightarrow$	ammonia	+	hydrogen chloride
$NH_4Cl(s)$	$\longrightarrow$	$NH_3(g)$	+	$HCl(g)$

C7.4.2–3 Equilibrium reactions

In a closed container, a reversible reaction reaches **equilibrium**. At equilibrium, all the reactants and products are present in the reaction mixture. Their amounts, or concentrations, do not change. The symbol $\rightleftharpoons$ shows that a reaction is at equilibrium.

H An equilibrium system is not static. For example, take the equilibrium reaction below:

$$NH_3(g) + HCl(g) \rightleftharpoons NH_4Cl(s)$$

All the time, ammonia and hydrogen chloride molecules are joining together to make ammonium chloride. This is the forward reaction. At the same time, ammonium chloride decomposes to make ammonia and hydrogen chloride. This is the backward reaction. The equilibrium is **dynamic**.

The forward and backward reactions happen at the same rate, so the amount of each substance in the equilibrium mixture does not change.

You can approach equilibrium from either direction – from the *product side* (right) of a reaction or the *reactant side* (left).

C7.4.4, C7.4.7 Fixing nitrogen

Plants need nitrogen compounds to grow. As the human population increases, more food is needed. Fertilisers containing nitrogen compounds increase crop yields.

Nitrogen is present in the air as nitrogen gas, N_2. But nitrogen gas is unreactive because its molecules contain very strong triple bonds. In **nitrogen fixation**, unreactive nitrogen is converted into nitrogen compounds that plants can use, including ammonia, nitrogen dioxide, and nitrates.

Nitrogen is fixed:

- at normal temperatures and pressures by bacteria that live in the root nodules of plants called legumes. Legumes include peas, beans, and clovers. The enzyme **nitrogenase** converts nitrogen to ammonia, NH_3.
- in the industrial **Haber process**, which produces huge amounts of ammonia every year. Most of the ammonia is used to make fertilisers, such as ammonium nitrate.

C7.4.5, C7.4a–d,g The Haber process

The feedstocks for the Haber process are:

- nitrogen – obtained from the air
- hydrogen – produced from a reaction between natural gas (mainly methane, CH_4) and steam.

Nitrogen and hydrogen form ammonia in a reversible reaction:

$$N_2(g) + 3H_2(g) \rightleftharpoons 2NH_3(g)$$

The gases in the reaction mixture can reach equilibrium if they are left in a closed container for enough time.

The amounts of chemicals in an equilibrium mixture change if the temperature or pressure is changed. **Le Chatelier's principle** helps to predict the effect on an equilibrium mixture of changing reaction conditions: *When conditions change, an equilibrium mixture responds so as to counteract the effect of the change.*

Changing pressure

The equation shows four gas molecules on the left and two on the right:

$$N_2(g) + 3H_2(g) \rightleftharpoons 2NH_3(g)$$

Increasing the pressure shifts the equilibrium towards the right, increasing the yield of ammonia. Le Chatelier's principle predicts this – as the number of molecules decreases, so does the pressure.

Changing temperature

The forward reaction is exothermic:

$$N_2(g) + 3H_2(g) \rightleftharpoons 2NH_3(g) \qquad \Delta H = -92 \text{ kJ/mol}$$

Lowering the temperature increases the amount of ammonia in the equilibrium mixture. Le Chatelier's principle predicts this. When the temperature is lower, more of the heat energy given out by the exothermic reaction can be absorbed.

Haber process conditions

In the Haber process the gases do not stay in the reactor for long enough to reach equilibrium.

The conditions chosen for the reaction are:
- pressure = 200 times atmospheric pressure
- temperature = 450 °C
- iron catalyst.

Under these conditions, the yield of ammonia is about 15%. This is increased by recycling unreacted hydrogen and nitrogen gas.

The conditions for the Haber process are a compromise. Chemists want to produce as much ammonia as possible as quickly as possible.
- Higher pressures increase the yield. But high-pressure equipment is expensive to build and run, and risky to operate.
- Low temperatures increase the yield of ammonia. But at low temperatures the reaction is slow. The temperature chosen, 450 °C, is a compromise between the need to maximise both yield and rate.

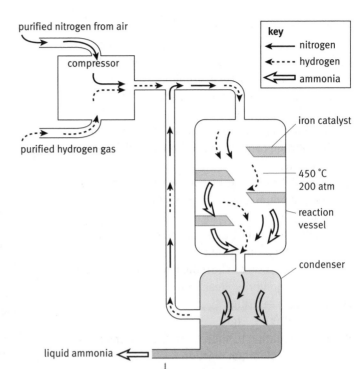

purified nitrogen from air

compressor

purified hydrogen gas

key
→ nitrogen
⤍ hydrogen
⇐ ammonia

iron catalyst

450 °C
200 atm

reaction vessel

condenser

liquid ammonia ⇐

This diagram summarises the Haber process.

C7.4.7e,f, C7.4.8 Catalysing the Haber process

In the Haber process, an iron catalyst increases the reaction rate. Chemists are searching for better catalysts, to make the process more efficient. For example, ruthenium catalysts increase the yield to 20% even at lower pressures.

Chemists are also developing new catalysts that mimic nitrogen-fixing enzymes. These will increase the efficiency of the Haber process, and allow it to work well at lower temperatures and pressures. They will reduce the high energy costs of the process.

C7.4.9–10 Impacts of nitrogen fertilisers

Nitrate fertilisers run off the land into rivers and lakes. Here, they make algae grow quickly. The algae damage ecosystems.

Nitrates harm health if they get into drinking water.

Use extra paper to answer these questions if you need to.

1 Write **T** next to each statement that is true for a system at equilibrium. Write **F** next to each statement that is false for a system at equilibrium.
   a Each reactant and product is present in an equilibrium mixture.
   b At equilibrium, the amounts of products change.
   c Equilibrium can be approached from the product side only.
   d An equilibrium mixture contains reactants only.
   e At equilibrium, amounts of reactants do not change.
   f Equilibrium can be approached from the reactant side or the product side.

2 Write one equation to show the two reactions below as a single equilibrium reaction.
   $$H_2(g) + I_2(g) \longrightarrow 2HI(g)$$
   $$2HI(g) \longrightarrow H_2(g) + I_2(g)$$

3 Write one equation to show the two reactions below as a single equilibrium reaction.
   $$CaCO_3(s) \longrightarrow CaO(s) + CO_2(g)$$
   $$CaO(s) + CO_2(g) \longrightarrow CaCO_3(s)$$

4 Use words from the box to complete the sentences below. Each word may be used once, more than once, or not at all.

peas	hydrogen	enzymes	air
beans	fixation	ammonia	grass

Plants need nitrogen compounds to grow. Nitrogen gas is present in the _____. Most plants cannot use this nitrogen as it is. Bacteria that live in the roots of plants such as _____ and _____ have _____ that convert nitrogen gas into _____. This process is called nitrogen _____.

5 Draw straight lines to match each feedstock for the Haber process with the one or more substances it is made from.

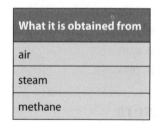

Feedstock
hydrogen
nitrogen

What it is obtained from
air
steam
methane

6 Describe the relationship shown by the graph below.

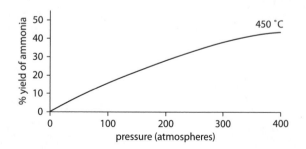

7 Write down the conditions that are usually chosen for the Haber process in the table below.

Temperature (°C)	
Pressure (atmospheres)	
Catalyst	

8 Write **I** next to changes that increase the yield of ammonia in the Haber process. Write **D** next to changes that decrease the yield of ammonia.
   a increasing the pressure
   b using a catalyst
   c increasing the temperature
   d recycling unreacted hydrogen

9 Making certain changes to the Haber process could make it more sustainable. Draw lines to match each possible change to the best reason for making it.

Change	Reason
obtain hydrogen from water only instead of from methane and steam	to reduce greenhouse gas emissions
use a more efficient catalyst	to reduce energy use during production
supply energy from hydroelectric plants instead of from burning fossil fuels	to leave stocks of non-renewable resources for future generations

10 Explain why chemists are interested in producing new catalysts to fix nitrogen that work in a similar way to natural enzymes.

11 Explain how the increasing use of nitrogen compound fertilisers damages the environment.

H 12 The graph shows the relationship between temperature and yield for the Haber process.

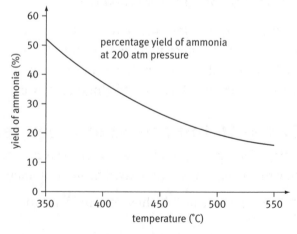

   a Describe the relationship shown by the graph.
   b Explain why a temperature of 450 °C is chosen for the Haber process.

13 Use the equation below to help you explain dynamic equilibrium.
   $$2NO_2(g) \rightleftharpoons N_2O_4(g)$$

1 Citric acid is a weak acid.
It is found in lemons and oranges.

Martha adds citric acid crystals to water.

The system soon reaches dynamic equilibrium.

The equation below represents the equilibrium reaction.

citric acid + water $\rightleftharpoons$ citrate ions + hydrogen ions

a Write a word equation for the forward reaction.

_____ [1]

b Name the chemicals present in the equilibrium mixture.

_____

_____ [1]

Ⓗ c Which of the following statements about the equilibrium reaction are true?

Tick **three** boxes to show the best answers.

In solution, only some citric acid is ionised. ☐

In the equilibrium mixture, citrate ions and hydrogen ions are reacting to make citric acid and water. ☐

The backward reaction is faster than the forward reaction. ☐

In the equilibrium mixture, citric acid and water are reacting to make citrate ions and hydrogen ions. ☐

The forward reaction is faster than the backward reaction. ☐

In solution, all the citric acid is ionised. ☐

[3]

Total [5]

2 The chemical industry uses the Haber process to manufacture huge amounts of ammonia.

a The Haber process is based on the reaction below.

$N_2(g) + 3H_2(g) \rightleftharpoons 2NH_3(g)$

i What does the symbol $\rightleftharpoons$ tell you about the reaction?

_____ [1]

ii Explain why, in the Haber process, the gases shown in the equation do not reach equilibrium.

_____ [1]

b The Haber process takes place in apparatus like that opposite.

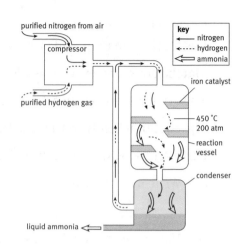

purified nitrogen from air

key
← nitrogen
←---- hydrogen
⇐ ammonia

compressor

iron catalyst

purified hydrogen gas

450 °C
200 atm

reaction vessel

condenser

liquid ammonia ⇐

    i Explain why unreacted nitrogen and hydrogen are recycled.

    _____ [1]

    ii Give the purpose of the catalyst.

    _____ [1]

c The graph below right shows the relationship between temperature and yield of ammonia.

    i Use the graph to describe the relationship between temperature and yield of ammonia.

    _____

    _____ [2]

    ii Use the equation and data below to give a reason for the relationship you described in part (i).

$$N_2(g) + 3H_2(g) \rightleftharpoons 2NH_3(g) \quad \Delta H = -92 \text{ kJ/mol}$$

    _____

    _____ [2]

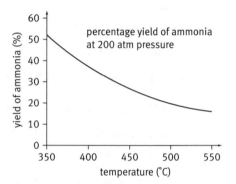

percentage yield of ammonia at 200 atm pressure

yield of ammonia (%) / temperature (°C)

d The yield of ammonia increases as pressure increases. Explain why.

_____

_____ [2]

Total [10]

⊕ Going for the highest grades

3 The chemical industry produces huge amounts of sulfur trioxide gas each year.

The equation for the reaction is given below:

$$2SO_2(g) + O_2(g) \rightleftharpoons 2SO_3(g)$$

Use the equation for the reaction, the graph opposite, and your own knowledge, to suggest why the reaction is normally carried out at a temperature of 450 °C and at atmospheric pressure.

The quality of written communication will be assessed in your answer to this question.

Write your answer on separate paper or in your exercise book.

Total [6]

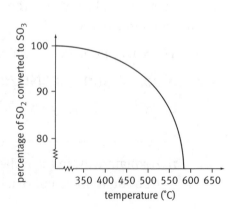

percentage of SO₂ converted to SO₃ / temperature (°C)

1   The diagram shows the paper chromatogram of a green felt-
    tip pen.

    Use the words and phrases in the box to annotate the diagram.
    Use each word or phrase once, more than once, or not at all.
    Write one word or phrase in each small box on the diagram.

solvent front	mobile phase
this dye moves up the paper faster	an aqueous solvent
this dye moves up the paper slower	a non-aqueous solvent
stationary phase	equilibrium lies towards the left
equilibrium lies towards the right	lid

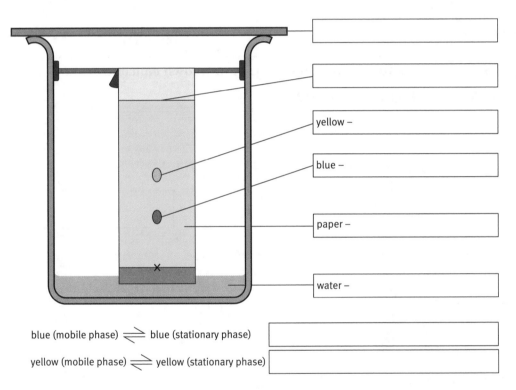

yellow –

blue –

paper –

water –

blue (mobile phase) ⇌ blue (stationary phase)

yellow (mobile phase) ⇌ yellow (stationary phase)

2   Use the gas chromatogram to answer the questions below.

    a   Which compound has the shortest retention
        time? _____

    b   Which compound passes through the column most
        slowly? _____

    c   Which peak is largest? _____

    d   Which compound makes up the highest proportion
        of the mixture? _____

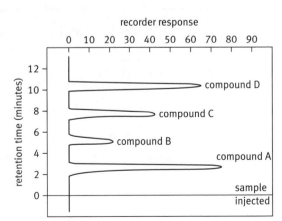

3   Write each letter in the correct section of the Venn
    diagram on the next page.

    A   The stationary phase is an absorbent solid supported on
        glass or plastic.

    B   The mobile phase is a liquid solvent.

    C   The mobile phase is helium gas.

D The stationary phase is absorbent paper.

E This method can be quantitative.

F This method gives qualitative information only.

G This method separates mixtures.

H Analysts use this method to measure blood-alcohol levels in drink-driving suspects.

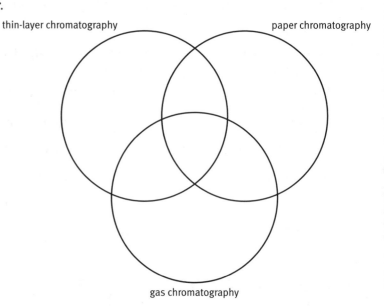

thin-layer chromatography   paper chromatography

gas chromatography

4 Nathan wants to find the concentration of a solution of hydrochloric acid. He sets up the apparatus below.

a Write one word or phrase in each small box below on the diagram.

pipette	burette	sodium hydroxide solution of known concentration
indicator	flask	hydrochloric acid

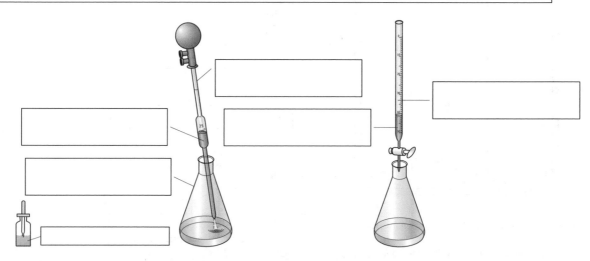

b Nathan's results are in the table.

Titration number	Rough	1	2	3	4
Final burette reading in cm³	49.2	48.7	50.0	37.6	26.0
Initial burette reading in cm³	23.1	23.7	24.9	12.7	1.0
Volume of acid used in cm³	26.1	25.0	25.1	24.9	25.0

i Calculate the average volume of hydrochloric acid added from the burette. Ignore the rough titration.

_____

_____

ii Write down the range of the results. Ignore the rough titration.

_____

_____

C7.5.1–4 Analytical procedures

Chemists do **qualitative** analysis to identify chemicals in a sample. They do **quantitative** analysis to find out amounts of chemicals.

Chemists use standard methods to collect, store, and prepare samples for analysis. This means that we can trust and compare analysis results.

- Collection – the sample must represent the bulk of the material.
- Storage – the sample must not 'go off' or be contaminated.
- Preparation – the sample is often dissolved in solution.

C7.5.5–15 Chromatography

In chromatography a **mobile phase** moves through a **stationary phase**. A chemist adds a sample of a mixture to the stationary phase. The mobile phase flows along. This makes the chemicals in the sample move through the stationary phase. Each chemical moves at a different speed. So the chemicals separate.

The chemicals move at different speeds because their molecules distribute themselves between the mobile and stationary phases. For each chemical there is a dynamic equilibrium between the two phases. So for a mixture of two chemicals, X and Y, there are two dynamic equilibria, shown opposite.

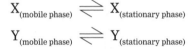

$$X_{(mobile\ phase)} \rightleftharpoons X_{(stationary\ phase)}$$
$$Y_{(mobile\ phase)} \rightleftharpoons Y_{(stationary\ phase)}$$

The position of equilibrium determines how quickly a chemical moves. If the equilibrium lies towards the mobile phase, the chemical moves quickly.

Paper and thin-layer chromatography

In **paper chromatography**, the stationary phase is chromatography paper. In **thin-layer chromatography** (TLC), the stationary phase is an absorbent solid supported on glass or plastic.

In both paper and thin-layer chromatography the mobile phase is a solvent. It may be an **aqueous solvent** (water) or a **non-aqueous solvent** (one with no water in it).

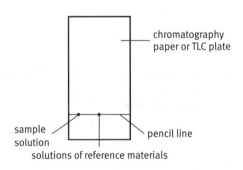

To analyse a sample by paper chromatography or TLC:

1 *Prepare the paper or slide*
- Draw a pencil line and add a drop of the sample.
- Put spots of **reference material** solutions on the line. These are substances you suspect might be in the unknown mixture.

2 *Run the chromatogram*
- Set up this apparatus.
- When the solvent gets near the top, take out the paper or plate.
- Mark the **solvent front**.

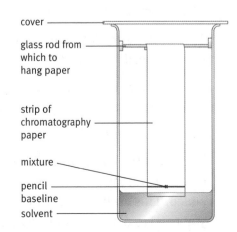

3 *Locate the separated chemicals*
   - Use pencil to draw around *coloured* spots.
   - Use a **locating agent** to find *colourless* spots.

4 *Interpret the chromatogram*
   - Identify spots by comparing them with reference material spots
   - *Or* calculate a chemical's **retardation factor** (R_f).

$$R_f = \frac{\text{distance moved by chemical}}{\text{distance moved by solvent}} = \frac{y}{x}$$

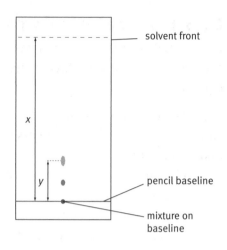

Gas chromatography

Gas chromatography (GC) separates mixtures and detects chemicals. Chemists use GC qualitatively and quantitatively.

The stationary phase is a thin film of liquid on the surface of a powder. The powder is packed into a long, thin tube. The mobile phase – or **carrier gas** – is helium.

To analyse a sample by GC:

1 *Use the oven* to keep the column at the correct temperature.

2 *Turn on the carrier gas* and adjust its flow rate.

3 *Inject the sample where the column enters the oven*. The chemicals in the sample become gases and mix with the carrier gas. The mixture moves through the column to the detector.

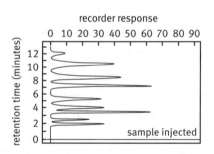

4 *Interpret the chromatogram:*
   - Each peak represents one compound.
   - The position of a peak shows its **retention time** (the time for it to pass through the column).
   - Peak heights show relative amounts of substances.

C7.5.16, C7.5.17, C7.5.21 Quantitative analysis by titration

Chemists use titrations to measure concentration, to check purity, and to find out the amounts of chemicals that react together.

To use an acid–base titration to find the concentration of a sodium hydroxide solution:

1 *Make up a standard solution of acid* and pour it into a burette. Read the burette scale.

2 *Use a pipette* to measure accurately a fixed volume of sodium hydroxide solution. Put it in a conical flask.

3 *Add a few drops of indicator*.

4 *Add acid from the burette*. Stop when the indicator changes colour. Read the burette scale. Calculate the amount of acid added. This **rough titration** gives an idea of how much acid you need to neutralise the sodium hydroxide, or reach the **end point**.

5 *Repeat the procedure*, but add the acid one drop at a time when you near the end point. Continue until you have three consistent values for the volume of acid.

6 *Calculate the mean and range* for your results.

C7.5.18 Preparing a standard solution

This is how to prepare a solution of sodium carbonate with an accurately known concentration.

1 Accurately weigh the sodium carbonate.

2 Dissolve the solute in a small amount of solvent, warming it if necessary.

3 Transfer the sodium carbonate solution to a graduated flask.

4 Rinse all the solution into the flask with more solvent.

5 Add solvent drop by drop to make up the volume to the mark on the flask.

6 Stopper and shake the flask.

⊕ C7.5.19, C7.5.20, C7.5.23 Titration calculations

1 *What is the concentration of sodium chloride solution made by dissolving 5 g of solid in water and making the volume up to 500 cm³?*

$$\text{concentration} = \frac{\text{mass of solute}}{\text{volume of solution}} = \frac{5\,\text{g}}{500\,\text{cm}^3} = 0.01\,\text{g/cm}^3$$

Multiply by 1000 to find the mass in 1 dm³: $0.01\,\text{g/cm}^3 \times 1000\,\text{cm}^3/\text{dm}^3 = \textbf{10 g/dm}^3$

2 *What mass of solute is in a 50 cm³ sample of a solution of hydrochloric acid of concentration 36 g/dm³?*

$$\text{mass of solute} = \text{concentration} \times \text{volume} = 36\,\text{g/dm}^3 \times \frac{50\,\text{dm}^3}{1000} = \textbf{1.8 g}$$

3 *Zac uses 25.0 cm³ of hydrochloric acid of concentration 3.65 g/dm³ to neutralise 25 cm³ of sodium hydroxide solution. What is the concentration of the sodium hydroxide solution?*

$$\text{HCl(aq)} + \text{NaOH(aq)} \longrightarrow \text{NaCl(aq)} + \text{H}_2\text{O(l)}$$

Reacting masses: 36.5 g 40 g

In 25.0 cm³ of the hydrochloric acid solution there is: $\frac{25\,\text{cm}^3}{1000\,\text{cm}^3} \times 3.65\,\text{g} = 0.0913\,\text{g}$ of acid

If 36.5 g of hydrochloric acid reacts with 40 g of sodium hydroxide, then 0.0913 g of acid reacts with:

$$\frac{0.0913\,\text{g}}{36.5\,\text{g}} \times 40\,\text{g} = 0.100\,\text{g of sodium hydroxide}$$

This mass of sodium hydroxide is in 25 cm³ (0.025 dm³) of solution. So the concentration of sodium hydroxide is:

$$\frac{\text{mass of solute}}{\text{volume of solution}} = \frac{0.100\,\text{g}}{0.025\,\text{dm}^3} = \textbf{4.00 g/dm}^3$$

Use extra paper to answer these questions if you need to.

1 Draw straight lines to match each process an analyst carries out with one reason.

Process	Reason
qualitative analysis	to prevent samples being contaminated or tampered with
quantitative analysis	to find out the amounts of the chemicals in a sample
store samples safely	to be able to compare samples to each other or to standard results
follow a standard procedure to collect samples	to find out which chemicals are in a sample

2 The stages below describe how chemists use gas chromatography (GC) instruments. They are in the wrong order.

A Turn on the carrier gas and adjust its pressure to get the correct flow rate.

B Wait while the chemicals in the sample turn to gases, mix with the carrier gas, and pass through the column.

C Turn on the oven and wait until the column reaches the correct temperature.

D Use a syringe to inject a sample at the start of the column.

E Interpret the printout.

Fill in the boxes to show the correct order. The first one has been done for you.

C				

3 Some of the stages of thin-layer chromatography (TLC) are described below. Give a reason for each stage.

a Dissolve the sample in a solvent.

b Use a pencil (not pen) to draw a line near the bottom of the plate.

c Cover the beaker with a piece of glass.

d Mark the final solvent front on the chromatogram.

e Spray the chromatogram with a locating agent chemical.

4 On the diagram here, calculate the R_f values for spots X and Y.

5 A student writes out the steps for making up a standard solution of sodium hydroxide. She makes one mistake in each step. Write a corrected version of each step.

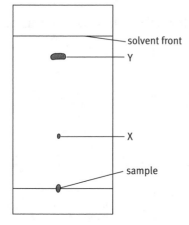

solvent front
Y
X
sample

a Roughly weigh 1.0 g of sodium hydroxide.

b Dissolve the sodium hydroxide in a small volume of tap water in a beaker.

c Transfer the solution to a 250 cm³ measuring cylinder.

d Rinse all of the solution from the beaker using more tap water.

e Add more water up to the 250 cm³ mark on the measuring cylinder.

f Place your hand over the top of the measuring cylinder and shake it.

6 Emma and Jess each do a set of six titrations to find the concentration of a solution of hydrochloric acid. Their results are in the table.

	Volume of sodium hydroxide solution (cm³)					
	Run 1	Run 2	Run 3	Run 4	Run 5	Run 6
Emma	24.2	24.4	25.0	25.8	25.8	25.6
Jess	25.2	24.8	25.2	25.0	24.8	25.0

a Which set of results will give a better estimate for the concentration of the acid? Explain why.

b There is not a significant difference between Emma and Jess's results. Explain how the data in the table shows this.

7 Calculate the concentrations of the solutions in the table. Give your answers in g/dm³.

Solute		Volume of solution (cm³)
Name	Mass of solid in solution (g)	
copper sulfate	4	1000
sodium chloride	15	500
magnesium sulfate	30	100
zinc bromide	40	250

8 Calculate the mass of solute in each of the solutions below.

a 1 dm³ of sodium hydroxide solution of concentration 40 g/dm³

b 250 cm³ of nitric acid of concentration 6 g/dm³

c 100 cm³ of potassium chloride solution of concentration 30 g/dm³

d 500 cm³ of sodium iodide solution of concentration 100 g/dm³

9 a Harriet uses 21.0 cm³ of hydrochloric acid of concentration 3.65 g/dm³ to neutralise 25.0 cm³ of potassium hydroxide solution. Calculate the concentration of the potassium hydroxide solution. The equation for the reaction is below.

$HCl(aq) + KOH(aq) \longrightarrow KCl(aq) + H_2O(l)$

b Ryan uses 48.0 cm³ of sodium hydroxide of concentration 4.0 g/dm³ to neutralise 25.0 cm³ of nitric acid. Calculate the concentration of the nitric acid. The equation for the reaction is below.

$HNO_3(aq) + NaOH(aq) \longrightarrow NaNO_3(aq) + H_2O(l)$

1 | Ron drove through a red traffic light. A police officer stopped him and took him to the police station.
At the police station, a doctor cleaned Ron's arm with an alcohol-free wipe. Then the doctor took a blood sample from Ron's arm. He divided the sample into two. He sealed and labelled each sample. The doctor gave one sample to the police officer and one to Ron. The police officer sent her sample to a laboratory for analysis. Ron took his sample home and put it in the fridge.

At the laboratory, a chemist looked carefully at the seals and labels on the sample. She injected part of the sample into a gas chromatography (GC) instrument, and obtained a printout.

a   i   Suggest why the doctor used an alcohol-free wipe to clean Ron's arm.

_____ [1]

   ii   Suggest why the doctor labelled the samples carefully.

_____ [1]

   iii Suggest why the chemist looked carefully at the seals on the sample.

_____ [1]

   iv   Suggest why Ron kept his sample in the fridge.

_____ [1]

b   Below is the GC printout for the sample of Ron's blood that was sent to the lab.

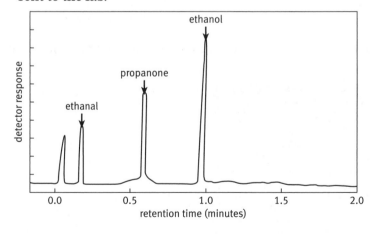

i   Estimate the retention time for the alcohol (ethanol) peak.

_____ [1]

ii   How many different chemicals are shown on the printout?

_____ [1]

c   Ron thinks the results of the analysis are incorrect.

He wants to check the results.

He takes the sample the doctor gave him to another laboratory.

A chemist at this laboratory injects part of the sample into a GC instrument.

He obtains the printout below.

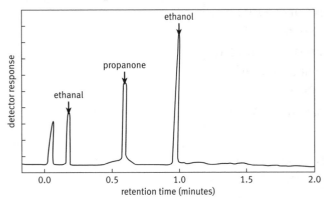

Does this printout support Ron's belief that the results of the first analysis are incorrect?

_____

Give a reason for your answer. _____ [1]

Total [7]

2 The chromatograms below show the coloured compounds in two fruit juices.

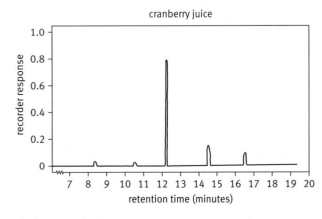 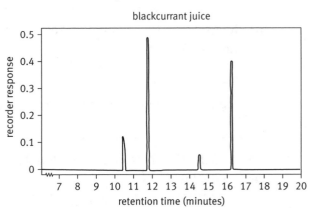

The table shows the retention times of some of the coloured compounds and each has a letter to identify it.

Compound	Letter to identify compound	Retention time (min)
delphinidin 3-galactoside	A	8.3
delphinidin 3-glucoside	B	10.5
delphinidin 3-rutinoside	C	11.8
cyanidin 3-galactoside	D	12.2
cyanidin 3-glucoside	E	14.5
cyanidin 3-rutinoside	F	16.3
cyanidin 3-arabinoside	G	16.5

Use the chromatograms, and data from the table, to compare the compounds in the two fruit juices, and their relative amounts.

The quality of written communication will be assessed in your answer to this question. Write your answer on separate paper or in your exercise book.

Total [6]

Data: their importance and limitations

1

> It says here that the concentration of nitrogen dioxide in the air in London is more than it is anywhere else in Britain. I don't believe it. In my **opinion**, the idea's a load of rubbish!

2

> It's not just an idea. My newspaper has lots of **data to justify the statement**. Scientists have measured the nitrogen dioxide concentration in many places.

3

> OK. But how do we know their data are **any good**? Do we know **how close to the true value** their measurements are?

4

> Good question. Maybe the measuring instruments were faulty. And even if all the scientists used the same instrument, they could have used it in different ways and got different results. And, of course, the concentration of nitrogen dioxide in any one place changes all the time!

5

> So you're saying the data are not accurate?

6

> No. In each place, the scientists took **many measurements**.

7

Why did they bother?

8

*If they just made one measurement, they could not have been sure it was accurate. So they made many measurements. Then they calculated the mean. The **mean is a good estimate of the true value**.*

9

Here's another problem: some of the data could be very different from the true value. Maybe something odd happened while they were taking measurements, or someone made a mistake.

10

*Exactly. One or two of the readings might have been very different from most of the others – they're called **outliers**.*

11

Where possible, scientists check an outlier to see if it is correct or not. If it is wrong, scientists discard it before calculating the mean.

12

If scientists cannot check an outlier, but have good reason to think it is inaccurate, they may discard it. But if there is no reason to discard an outlier, they treat it as data.

Data: their importance and limitations

1 Solve the clues to fill in the grid on the opposite page.
You will need the data in the table to help you fill in some of
the words.

Metal	Melting point (°C)	Density (g/cm³)
gold	1063	19.3
lead	327	11.3
iron	1535	7.9
silver	961	10.5
cadmium	321	8.6
zinc	420	7.1

1 Scientists create explanations to account for . . .

2 A student measured the melting point of lead six times.
The highest value she obtained was 329 °C, and the lowest
value was 324 °C. So the . . . of the readings was 324–329 °C.

3 A measurement that lies well outside the range of the
others in a set of repeats is called an . . .

4 Data that are . . . do not vary much when you repeat the
measurements under the same conditions.

5 Of the metals in the table, iron has the highest . . .

6 Faulty measuring equipment leads to . . . data.

7 The best estimate of the value of a quantity is the . . . of
several repeat measurements.

8 Iron has a . . . density than zinc.

9 The mean of several repeat measurements is the best
estimate of the . . . value of a quantity.

10 To get the best estimate of the true value of a quantity,
take several . . . measurements and calculate the mean.

11 Of the metals in the table, zinc has the . . . density.

12 All the metals in the table are . . . at room temperature.

13 Scientists use data rather than . . . to justify an
explanation.

14 If the mean of a set of readings for one sample is within
the range of a set of readings for another sample, there
is . . . real difference between the true values of the two
samples.

15 A student measured the melting point of a metal in the
table several times. She calculated that the mean of her
measurements was 959 °C. She concluded that the metal
was probably . . .

1 d
2 a
3 t
4
a
5 l
6 i
7 m
8 i
9 t
10 a
11 t
12 i
13 o
14 n
15 s

2 Each table on the next page is designed to collect a set of data about an air pollutant, sulfur dioxide.

Give the letter or letters of the best table or tables in which to collect data to compare:

a the concentrations of a pollutant in the four seasons of the year ________

b the concentrations of a pollutant on weekdays and at the weekend ________

c the concentrations of a pollutant at different times of day ________

d the concentrations of a pollutant in two different cities ________

A

Place: Prague			
Date	Day	Time	Concentration of sulfur dioxide (µg/m³)
3 Jan	Mon	11.00	
6 April	Wed	11.00	
7 July	Thu	11.00	
7 Oct	Fri	11.00	

B

Place: Paris			
Date	Day	Time	Concentration of sulfur dioxide (µg/m³)
3 Jan	Mon	11.00	
5 Jan	Wed	11.00	
6 Jan	Thu	11.00	
9 Jan	Sun	11.00	

C

Place: Brussels			
Date	Day	Time	Concentration of sulfur dioxide (µg/m³)
3 Jan	Mon	11.00	
6 April	Wed	12.00	
7 July	Thu	10.00	
2 Oct	Sun	11.00	

D

Place: Leipzig			
Date	Day	Time	Concentration of sulfur dioxide (µg/m³)
7 July	Thu	08.00	
7 July	Thu	12.00	
7 July	Thu	17.00	
7 July	Thu	22.00	

1

Data: their importance and limitations

1 Ben needs to know the concentration of a solution of sodium hydroxide.

He places exactly 25.0 cm³ of sodium hydroxide solution in a conical flask.

He adds a few drops of indicator.

He titrates the mixture with hydrochloric acid.

His results are in the table below.

	Volume of hydrochloric acid required (cm³)			
	Rough	Run 1	Run 2	Run 3
Initial burette reading (cm³)	0.3	12.0	23.5	35.0
Final burette reading (cm³)	12.3	23.5	34.9	46.3
Volume of acid added (cm³)	12.0	11.5	11.4	

a   i   Suggest why Ben repeats the titration several times.

_____ [1]

ii   Suggest why the measurements of the volume of hydrochloric acid required have different values.

_____ [1]

b   Calculate the volume of acid added in run 3.

answer = _____ cm³   [1]

c   i   Identify the outlier in the set of data in the table.

_____ [1]

ii   Ben decides to discard the outlier. Suggest why.

_____ [1]

d   Estimate the range within which the true value for the volume of acid required probably lies.

_____ [1]

e   Calculate the mean volume of acid needed to exactly neutralise the sodium hydroxide solution.

answer = _____ cm³   [2]

Total [8]

2 Scientists measured the salt content of hamburgers from two restaurants.

They tested six hamburgers from each restaurant.

Their results are in the table.

Sample	Salt content (g)							
	1	2	3	4	5	6	Range	Mean
Restaurant A	0.9	1.1	1.0	1.3	0.5	1.2	0.5–1.3	1.0
Restaurant B	1.6	1.4	1.2	1.4	1.3	1.5		

a i Identify the outlier in the data from restaurant A.

Outlier is sample _____.   [1]

ii The scientists decided to include the outlier in the range and mean for restaurant A. Suggest why.

_____ [1]

b Work out the range and the mean for the samples from restaurant B.

Range = _____ to _____ g

Mean = _____ g   [2]

c The scientists conclude that there is a **real difference** between the salt content of the hamburgers from the two restaurants.

Explain how the data in the table and your answer to part **b** show this.

_____

_____ [1]

d A different group of scientists measured the salt content of hamburgers from restaurant B.

Their mean value was the same as that of the scientists whose results are in the table above.

What does this show?

Put a tick in **one** box next to the best answer below.

The data for restaurant B is repeatable. ☐

The two sets of data have the same range. ☐

The data for restaurant B is reproducible. ☐

The two sets of data are identical. ☐ [1]

Total [6]

3 Engineers from two factories tested the carbon dioxide emissions of two different cars of the same make and model.

Their data is in the table.

	Carbon dioxide emissions (g/km)					
Sample	1	2	3	4	5	6
Car A	153	158	159	163	157	158
Car B	156	164	160	163	160	157

Use the data to decide whether or not there is a real difference between the carbon dioxide emissions of the two cars.

Write down your decision.

Use data from the table to explain how and why you came to this decision.

The quality of written communication will be assessed in your answer to this question.

Write your answer on separate paper or in your exercise book.

Total [6]

4 A group of students in the Czech Republic studied data and devised this scientific explanation:

> There is a correlation between being exposed to air pollution and the percentage of sperm with damaged DNA.

a The students collected data on one pollutant, sulfur dioxide.

They measured its concentration at the same place every day for six months.
Their measurements were different every day.

Why were the measurements different every day?
Put ticks in the boxes next to three possible reasons.

Wind direction varied. ☐

A nearby coal-fired power station was running on some days, but not on others. ☐

One student used the measuring instrument incorrectly. ☐

A nearby nuclear power station was running on some days, but not on others. ☐ [1]

b Suggest what other data the students needed to collect to provide evidence for the explanation in the box.

_____

_____ [1]

Total [2]

Cause–effect explanations

China plans to reduce the power station emissions that cause acid rain. It says that acid rain damages its trees, crops, and buildings.

*OK. So one of the **outcomes** is that trees are damaged. But surely acid rain is not the only **factor** that affects the amount of damage?*

True. There are many other possible factors. Maybe diseases, pests, or even climate change damage China's trees. But there's lots of scientific evidence that acid rain is an important factor.

So changing this factor . . . reducing the amount of acid rain . . . could change the outcome? Fewer trees would be damaged?

Let's hope so!

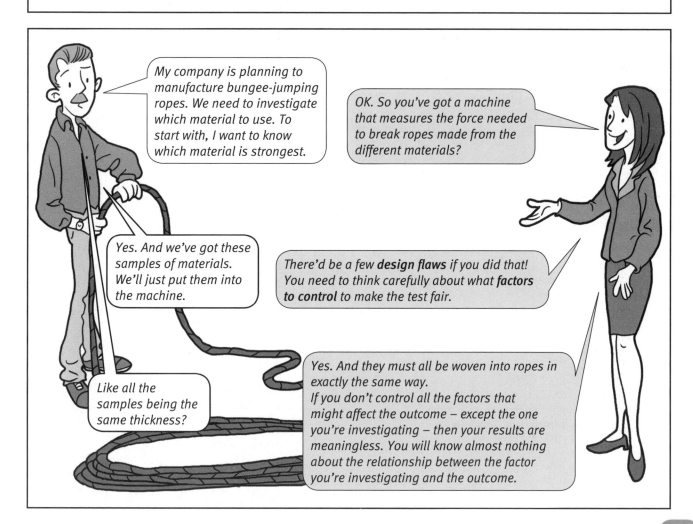

My company is planning to manufacture bungee-jumping ropes. We need to investigate which material to use. To start with, I want to know which material is strongest.

OK. So you've got a machine that measures the force needed to break ropes made from the different materials?

Yes. And we've got these samples of materials. We'll just put them into the machine.

*There'd be a few **design flaws** if you did that! You need to think carefully about what **factors to control** to make the test fair.*

Like all the samples being the same thickness?

Yes. And they must all be woven into ropes in exactly the same way.
If you don't control all the factors that might affect the outcome – except the one you're investigating – then your results are meaningless. You will know almost nothing about the relationship between the factor you're investigating and the outcome.

Cause–effect explanations

1 Circle the letters of the graphs that show a correlation between a factor and an outcome.

A

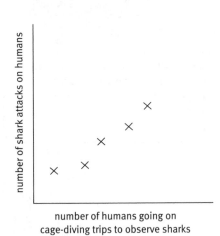

B

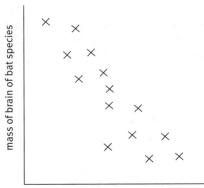

C

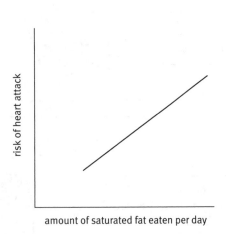

D

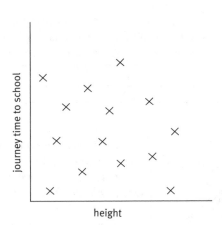

E

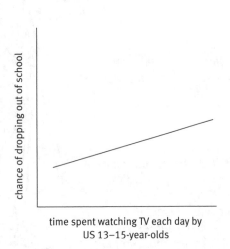

F

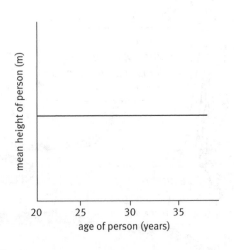

2   For each graph, chart, and table below, describe the
correlation between the factor and the outcome.

A

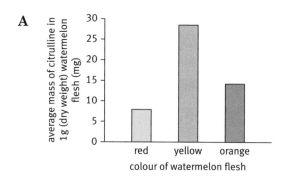

Correlation _____

_____

B

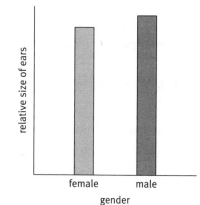

Correlation _____

_____

C

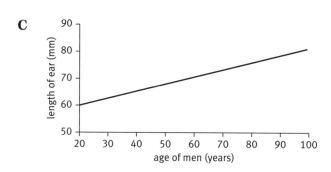

Correlation _____

_____

D

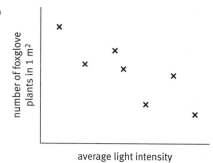

Correlation _____

_____

E

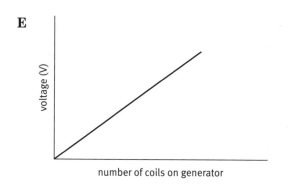

Correlation _____

_____

F

Temperature (°C)	Time to collect 100 cm³ gas in the reaction of magnesium with dilute hydrochloric acid (s)
10	120
20	60
30	31
40	16
50	7

Correlation _____

_____

Cause–effect explanations

1 A shampoo company claimed:

> In only 10 days, our antibreak shampoo gives up to 95% less hair breakage than other shampoos.

Scientists at the company tested 10 samples of hair, three times each.

a Identify the factor (input variable) and the outcome (outcome variable) in the investigation.

Factor: _____

Outcome: _____ [1]

b A school student decided to investigate the company's claims.
He used this apparatus to test hair strength.

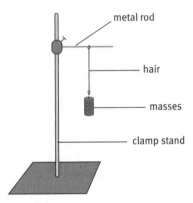

metal rod
hair
masses
clamp stand

i Give two factors the student must control.

_____

_____ [2]

ii Explain why the student must control these factors.

_____ [1]

c The student found that a greater weight was needed to break hair that had been washed with antibreak shampoo.

Explain why this correlation does not necessarily mean that the shampoo makes hair stronger. [1]

_____

_____

_____

Total [5]

2 In December 2005, there was a huge fire at an oil depot in southern England. Some of the chemicals in the smoke were:

> carbon monoxide, sulfur dioxide,
> nitrogen dioxide, solid carbon

a Tick one box next to each statement to show whether the statement is an outcome or a factor that may affect one or more of the outcomes. There are three factors and three outcomes. One has been done for you.

	Statement	Outcome	Factor
A	Smoke from the fire rose 3000 m above the ground.		
B	The atmospheric concentration of sulfur dioxide 3000 m over southern England increased.		
C	The atmospheric concentration of nitrogen dioxide near the ground over southern England changed very little.		
D	The smoke was trapped 3000 m above the ground.		✔
E	The atmospheric concentration of carbon monoxide over northern England did not change.		
F	Wind carried the smoke south.		

[2]

b Nitrogen dioxide gas is the only gas produced by the fire that increases the risk of asthma attacks.

Draw straight lines to match each comment with one conclusion.

Comment
1 When the concentration of carbon monoxide increases, there is no change in the number of asthma attacks.
2 More asthma sufferers had asthma attacks in the week after the fire than in a normal week. But not every asthma sufferer had an attack.
3 When the concentration of nitrogen dioxide gas in the air increases, more asthma sufferers have asthma attacks.
4 I know three people who had asthma attacks after the fire.

Conclusion
A There is a correlation between this factor and this outcome.
B These cases do not provide evidence for or against a correlation between the factor and the outcome.
C There is no correlation between this factor and this outcome.
D The factor increased the chance of the outcome but did not always lead to it.

[3]

Total [5]

3 The concentration of ozone in the upper atmosphere has decreased since 1960.

Ozone protects humans from the harmful effects of the Sun's ultraviolet radiation, like getting eye cataracts.

American scientists wanted to know if the 'thinning' ozone layer would lead to more people getting eye cataracts. They studied 2500 people.

Their results showed that **if the concentration of ozone decreases by 20%, the number of people with cataracts is likely to increase by 7%.**

a Tick the boxes next to the three statements that could be true.

The smaller the concentration of ozone in the upper atmosphere, the greater your chance of getting a cataract. ☐

There is a correlation between the concentration of ozone in the upper atmosphere and the number of people with cataracts. ☐

Wearing sunglasses that protect against ultraviolet radiation may reduce your chance of getting a cataract. ☐

If the concentration of ozone in the upper atmosphere increases, the number of people with cataracts is likely to increase. ☐ [1]

b A British scientist wants to find out if there is a correlation between the amount of exposure to the Sun's ultraviolet radiation and the risk of getting a cataract.

She decides to compare two groups of people. All the people in one sample have cataracts; all those in the other sample do not have cataracts.

i Draw a ring around the sample size that will give the most reliable conclusion.

**10   100   1000
as large as practically possible** [1]

ii The scientist makes sure that the two samples are matched for age.

Suggest two other factors that she must match.

_____

_____ [2]

Total [4]

> **Exam tip**
>
> Remember – just because there is a correlation between a factor and an outcome, it does not mean that the factor caused the outcome.

4 For an equilibrium reaction in which all the reactants and products are gases, an increase in pressure favours the reaction that produces fewer molecules, as shown by the equation for the reaction.

a The graphs show how the yield of a product changes with pressure for the equilibrium reactions below.

One of the graphs applies to two of the reactions.

A $H_2(g) + I_2(g) \rightleftharpoons 2HI(g)$

B $2SO_2(g) + O_2(g) \rightleftharpoons 2SO_3(g)$

C $N_2(g) + 3H_2(g) \rightleftharpoons 2NH_3(g)$

D $N_2O_4(g) \rightleftharpoons 2NO_2(g)$

Next to each graph, write the letter of one reaction above that the graph could represent.

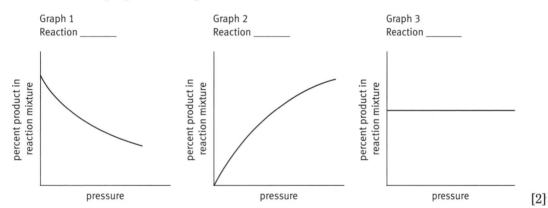

Graph 1
Reaction _____

Graph 2
Reaction _____

Graph 3
Reaction _____

[2]

b A chemist investigates the relationship between pressure and percentage of sulfur trioxide in the equilibrium mixture for the reaction below.

$2SO_2(g) + O_2(g) \rightleftharpoons 2SO_3(g)$

i Identify the outcome of the investigation.

_____ [1]

ii Identify **three** factors that might affect the outcome.

State which **two** of these factors the chemist must control, and give a reason for your choice.

_____

_____

_____ [2]

c The chemist investigates how the percentage of oxygen in the mixture that enters the reaction vessel affects the percentage of sulfur dioxide that is converted to sulfur trioxide for the reaction below.

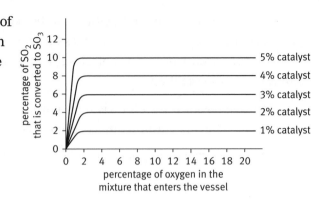

$2SO_2(g) + O_2(g) \rightleftharpoons 2SO_3(g)$

He repeats the investigation using different amounts of catalyst.

He obtains the graph on the right.

Describe fully the two correlations shown by the graph.

_____

_____

_____

_____

_____ [2]

d The graph opposite shows the correlation between temperature and the percentage of sulfur dioxide that is converted to sulfur trioxide.

Some students discuss the graph.

Carrie The graph shows that the higher the temperature, the greater the percentage of sulfur dioxide that is converted to sulfur trioxide.

Joe The only factor that could cause the change in the percentage of sulfur dioxide that is converted is temperature.

Leah To make the biggest amount of sulfur trioxide possible in the shortest time, the temperature should be as low as possible.

Ezekiel The graph shows that you can convert a higher percentage of sulfur dioxide if you use a catalyst.

Evaluate the comments of the students.

_____

_____

_____

_____

_____

_____

_____ [4]

Total [11]

Ideas about science

Developing scientific explanations

In the 1770s, French chemists Antoine Lavoisier and Maria-Anne Lavoisier were at work in their laboratory.

1

We know that air is a mixture of gases. But I want to know more. How do they behave? What are their properties?

Let's investigate.

2

We've been heating this mercury in its container of air for a whole day. Can you see any changes?

Yes, look! There are bits of red solid on its surface. Amazing!

3

Four days later...

There's even more red solid. Let's keep heating until there are no more changes.

4

After another week...

It hasn't changed for a few days now. Let's stop heating and make some measurements.

5

There's less air than there was. The volume has decreased from 50 units to about 42 units.

Interesting data.

6

What about the properties of the remaining air? Try putting a burning taper into it.

It's gone out. It's like putting a flame into water.

7

How about putting an animal into the leftover air. Can it still breathe?

No. It's suffocated, poor thing. But the data is fascinating.

8

Shall we heat the red solid, and collect anything it produces?

Good idea. It might give off a liquid or a gas...

9

There's a gas. We need to investigate the gas, collect more data.

Look at that taper – it's burning with dazzling splendour. And the charcoal. Its flame is so bright I can hardly look.

10

So we've got all this data. But how can we explain it? I need time to be creative, time to think.

11

I've got it! The air is made up of two gases. One of these supports breathing. I'll call it oxygen gas.

The other gas makes up more than 70% of the air. If animals breathe only this gas, they die.

12

That's a great explanation, Antoine. It accounts for all our data. And it explains things that, before, we did not think were linked. Who would have thought there is a connection between burning and breathing?

Those are excellent reasons for accepting the explanation.

13

I think so. But now we need to test the explanation further.

Can we use it to make predictions about new situations?

14

Let's heat phosphorus in a closed container of air.

I predict that the phosphorus will burn, and join with oxygen from the air, to make a new substance.

The volume of air will decrease by about one fifth because phosphorus will remove oxygen from it.

15

Your prediction was right, Antoine.

Now let me make another prediction to test the explanation.

If we heat some phosphorus in pure oxygen, and some in air, I predict that the burning will be faster in oxygen than in air.

16

Our predictions were right, Antoine. The burning reaction was much faster in oxygen. And, when we heated phosphorus in air, the volume of air decreased by one-fifth.

Wonderful. Our observations agree with our predictions. Now I'm even more confident in our explanation.

17

It still doesn't prove the explanation is correct, though, does it...

18

True, Maria-Anne. All we can do is make more predictions and collect more data.

The more they agree, the more convincing our explanation.

Developing scientific explanations

1 Write **T** next to the statements that are true and **F** next to the statements that are false.

    a   Scientific explanations simply summarise data.   ___

    b   If the data agree with an explanation, the explanation must be correct.   ___

    c   Developing a scientific explanation requires creative thought.   ___

    d   A scientific explanation must account for most, or all, of the data already known.   ___

    e   An explanation should always explain a range of phenomena that scientists didn't know were linked.   ___

    f   Scientists test explanations by comparing predictions based on them with data from observations or experiments.   ___

    g   If an observation agrees with a prediction that is based on an explanation, it proves that the explanation is correct.   ___

    h   If an observation does not agree with a prediction that is based on an explanation, it decreases confidence in the explanation.   ___

    i   If an observation does not agree with a prediction that is based on an explanation, then the observation must be wrong.   ___

    j   If an observation does not agree with a prediction that is based on an explanation, then the explanation must be wrong.   ___

Now write corrected versions of the **six** false statements on the lines below.

_____

_____

_____

_____

_____

_____

_____

_____

2 The statements below describe how a group of scientists made a compound that catalyses the formation of ammonia from nitrogen at atmospheric temperature and pressure.

Write the letter of each statement in an empty box on the flow chart.

One box needs two letters; the other three boxes need one letter each.

A Blue–green algae and other organisms use natural catalysts to convert nitrogen to ammonia.

B If we make a compound that has a single molybdenum atom at its centre, and supply protons and electrons, the compound will catalyse the formation of ammonia from nitrogen at atmospheric temperature and pressure.

C Catalysts in blue–green algae include atoms of the element molybdenum.

D We could supply protons and electrons to nitrogen in the presence of a molybdenum compound. We could use X-ray tests to find out if ammonia is formed.

E A possible explanation is that there is a molybdenum-containing compound that catalyses the formation of ammonia from nitrogen at atmospheric temperature and pressure.

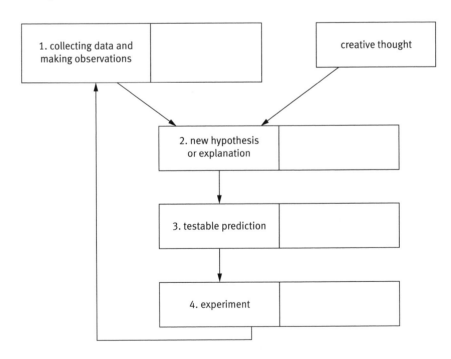

3 The statements below describe how a group of scientists have discovered how to produce butanol, an alcohol that can be used as a vehicle fuel, from waste paper.

Write the letter of each statement in an empty box on the flow chart.

One box needs four letters; the other three boxes need one letter each.

A TU-103 bacteria live in animal faeces (poo) and break down cellulose in the faeces to form butanol. Cellulose is the main material in plant cell walls.

B Paper is mainly cellulose.

C TU-103 bacteria can grow and reproduce in the presence of oxygen.

D Oxygen kills most butanol-producing bacteria.

E A possible explanation is that TU-103 bacteria make butanol from waste paper in the presence of oxygen.

F If we isolate and grow TU-103 bacteria and allow them to feed on waste paper, the bacteria will produce butanol fuel in the presence of oxygen.

G We could allow TU-103 bacteria to feed on bacteria in the presence of oxygen, and find out if butanol is made.

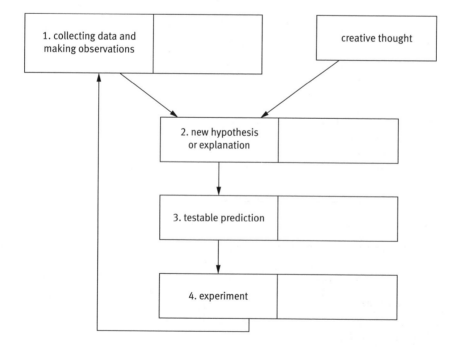

Developing scientific explanations

1 a In the 1920s, Russian scientist Alexander Oparin wondered how life began. He experimented with organic compounds in solution. He noticed that the chemicals sometimes organised themselves into droplets and layers.

Oparin came up with an explanation:

Explanation 1

Life began when the Earth had an atmosphere of methane, ammonia, hydrogen, and water vapour, and when there were many thunderstorms. In these conditions amino acids (organic compounds that are the basis of life) were formed.

Suggest why Oparin must have used creative thinking to help him develop this explanation.

_____

_____ [1]

b In the 1950s, another scientist, Stanley Miller, read about the work of Oparin.

Miller used Explanation 1 to make a prediction:

If methane, ammonia, hydrogen, and water vapour are circulated past electric sparks (to simulate lightning), amino acids will be formed.

Put ticks (✓) in the boxes next to the **two** statements that are true.

If the prediction is correct, we can be sure that the explanation is correct. ☐

If the prediction is correct, we can be more confident that the explanation is correct. ☐

If the prediction is wrong, we can be sure that the explanation is wrong. ☐

If the prediction is wrong, we can be less confident that the explanation is correct. ☐

[2]

c Miller did an experiment to test his prediction.

He mixed methane, ammonia, hydrogen, and water vapour.

He let the mixture circulate past electric sparks for a week.

He used paper chromatography to analyse the products of the reactions that took place.

His chromatogram looked like this:

Glycine, aspartic acid, and alanine are amino acids.

Does Explanation 1 account for the results on the chromatogram?

Explain your answer.

_____

_____ [2]

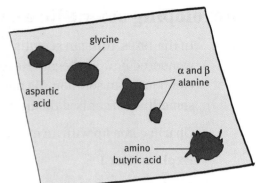

d Other scientists developed another explanation for how life began:

Explanation 2

Life began at deep sea vents. Hot chemicals bubble out of the vents.

Conditions are perfect for the chemical reactions that make amino acids, the organic compounds that are the basis for life.

i Scientists have made the observations **A–D** below.

Tick (✓) the two pieces of evidence that best support Explanation 2.

A Tiny organisms live in hot rock deep under the surface of the Earth. ☐

B At the temperature of deep sea vents, amino acids normally break down into smaller molecules. ☐

C Scientists have found ancient fossils near deep sea vents. ☐

D Most living things cannot survive at the high pressures near deep sea vents. ☐

[2]

ii Suggest how scientists might decide whether Explanation 1 or Explanation 2 is a better explanation for how life began.

_____

_____

_____

_____ [3]

Total [10]

The scientific community

Reporting scientific claims in journals and at conferences gives other scientists opportunities to critically evaluate their claims... and to build on their work.

5

1864. John Newlands speaks about his Law of Octaves at a London Chemical Society meeting.

I have arranged all 56 known elements in order of atomic mass. There is a pattern. Every eighth element has similar properties.

6

Why have you grouped copper with lithium, sodium, and potassium? Copper has very different properties.

Why arrange elements like a musical scale? You might as well have ordered them alphabetically!

This is such a new idea. It has not been evaluated by others. We cannot accept it.

7

1869. Dmitri Mendeleev creates the periodic table.

I arranged the elements in order of atomic mass, more or less. Elements with similar properties are grouped together.

I've left gaps for elements that have not yet been discovered.

8

1875. Paul-Emile Lecoq de Boisbaudran makes a discovery.

I've found one of Mendeleev's missing elements. I've called it gallium.

This evidence makes me more confident in Mendeleev's claims.

9

1879. Lars Nilson discovers another missing element.

I've found the element between calcium and titanium. Its name is scandium.

I am very confident that Mendeleev's ideas are correct.

The scientific community

1 The statements below describe some of the steps by which scientists may accept a new scientific claim or explanation.

Write each statement in a sensible place on the flow chart below.

A Many other scientists read the paper. Some may try to reproduce its findings.

B A small number of other scientists read the paper to check the methods and claims, and to spot any mistakes.

C He makes a claim or creates a new explanation.

D The scientist makes corrections to his paper.

E He repeats the investigation to check that he can replicate his own findings.

F Other scientists are sceptical about the new claim or explanation.

G The paper is published in a scientific journal, in print, and online.

H Other scientists are more likely to accept the new claim or explanation.

4

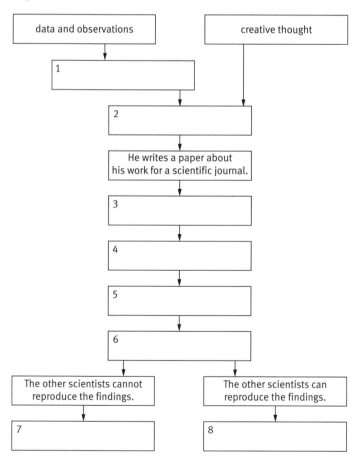

The scientific community

1 When Mendeleev created the periodic table, he left gaps for elements that had not yet been discovered.

By 1900, nearly all the gaps had been filled.

But two elements were still missing – those under manganese with atomic numbers of 43 and 75.

a   In 1909, Japanese scientist Masataka Ogawa extracted a tiny amount of an unknown chemical from a mineral.

He claimed he had found element 43, and called it nipponium.

He published his findings in a scientific journal.

A second scientist tried to reproduce the findings of Masataka Ogawa. He could not extract the new element from the mineral.

How might the second scientist's result have influenced other scientists? Tick the **two** best answers.

The result made other scientists less likely to accept Masataka Ogawa's claim. ☐

The result shows that Masataka Ogawa definitely did not find a new element. ☐

The result made other scientists more likely to accept Masataka Ogawa's claim. ☐

The result made other scientists more likely to question Masataka Ogawa's claim. ☐   [2]

b   Read the article opposite about the discovery of the element rhenium.

> In the early 1920s Ida Tacke and William Noddack extracted 1 g of a new chemical from 660 kg of an ore. The scientists thought it might be the missing element of atomic number 75.
>
> The scientists asked another scientist, Otto Berg, to look at the spectrum of the new chemical. He found some new lines in the spectrum. He explained that the new lines were from the new element.
>
> Ida Tacke announced the discovery of the new element at a conference in 1925.

i   Suggest why Ida Tacke and William Noddack asked Otto Berg to look at the spectrum of the new chemical before telling other scientists about it at the conference.

_____

_____ [2]

ii In 1925 Ida Tacke and William Noddack also announced that they had discovered missing element 43.

At the time, other scientists could not reproduce their findings. They thought that Ida and William were wrong.

In 1999 another scientist repeated the work of Ida and William, and obtained a similar result.

Suggest how the result of the 1999 scientist might influence other scientists' opinions on the work of Ida and William.

_____ [1]

Total [5]

Ideas about science

Risk

1 *TV – Local news*

A company is planning to build an ammonia plant on the outskirts of the city.

A company spokesperson says the new facility is needed to meet the growing demand for fertilisers such as ammonium nitrate.

2 I'm moving if they make ammonia here. It's too dangerous. What if the plant explodes?

The **chance** of it exploding is tiny. Hundreds of chemical plants make ammonia. Few have accidents of any kind.

3 But you need to think of the **consequences** if it did explode. Ammonia is toxic. And corrosive.

Yes. I heard of ammonia leaking in a chicken-processing factory. Workers ended up in hospital.

4 You're over-reacting. Like I said, the **statistically estimated** risk of an explosion is tiny.

Are you overestimating the risks because ammonia gas is invisible? Or is it because you're not used to it?

5 All I know is I don't want it. I've not chosen to live near an ammonia plant. And if it did explode, the health effects could last for years.

6 But we need ammonia to make fertilisers. Surely the benefits of fertilisers outweigh the risks linked to making ammonia.

And there are laws about these things. Safety organisations have worked out acceptable risk levels for ammonia plants. Let them make ammonia!

5

Risk

1 Use the clues to fill in the grid.

1 Everything we do carries a risk of accident or h . . .

2 We can assess the size of a risk by measuring the c . . . of it happening in a large sample over a certain time.

3 New vaccines are an example of a scientific a . . . that brings with it new risks.

4 Radioactive materials emit i . . . radiation all the time.

5 The chance of a nuclear power station exploding is small. The c . . . of this happening would be devastating.

6 Governments or public bodies may have to a . . . what level of risk is acceptable in a particular situation.

7 Some decisions about risk may be c . . ., especially if those most at risk are not those who benefit.

8 Sometimes people think the size of a risk is bigger than it really is. Their perception of the size of the risk is greater than the s . . . calculated risk.

9 Nuclear power stations emit less carbon dioxide than coal-fired power stations. Some people think that this b . . . is worth the risk of building nuclear power stations.

10 Many people think that the size of the risk of flying in an aeroplane is greater than it really is. They p . . . that flying is risky because they don't fly very often.

11 It is impossible to reduce risk to zero. So people must decide what level of risk is a . . .

2 Draw a line to link two words on the circle.
Write a sentence on the line saying how the two words
are connected.
Repeat for as many pairs as you can.

risk

safe   benefits

chance   consequences

5

balance   unfamiliar

scientific advances   long-lasting effects

statistically estimated risk   perceived risk

controversial

Risk

1 Read the information in the box.

> There is strong evidence that eating too much salt is linked to high blood pressure. High blood pressure is the main cause of strokes, and a major cause of heart disease. Diets high in salt may also cause stomach cancer, and make asthma symptoms worse.
>
> The table shows the amount of salt in some foods.
>
> | Food | Mass of salt in 100 g of food (g) |
> |---|---|
> | bread | 1.5 |
> | spread | 1.4 |
> | porridge oats | trace |
> | rice flakes breakfast cereal | 1.2 |
> | milk | 0.04 |

  a The recommended maximum daily salt intake for an adult is 6 g a day.

      Use the table above to work out the mass of bread that contains this mass of salt.

Answer = _____ g   [2]

  b **i** Oliver has high blood pressure.

          His doctor tells him to cut down on salt to reduce his risk of heart disease.

          Oliver wants to choose a low-salt breakfast to have every day. Should he choose 75 g of bread with 10 g of spread **or** 50 g of rice flakes with 200 g milk?

          Show your working to help explain your answer.   [4]

      ii Explain why eating porridge every day (made from oats and milk) would further reduce the risk of Oliver getting heart disease.

          _____ [1]

  c Many people like salty food. Even so, some food manufacturers have reduced the amount of salt in the foods they make. Suggest why.

      _____ [1]

  d The British government is working with the food industry to reduce salt levels in food, but has not made food with high levels of salt illegal. Suggest why.

      _____

      _____ [1]

Total [9]

Making decisions about science and technology

1

Don't buy a new phone. You don't need it. You've only had the old one a year.

New phones cost the Earth – and lives.

2

What do you mean? I couldn't live without my phone. And they don't just benefit people in rich countries. My uncle in Tanzania says his mobile phone has changed his life. His business makes much more money now.

3

Of course. Mobile phones are a perfect example of a science-based technology improving the quality of life.

But there are costs too.

4

Costs? What costs? How can a phone cause harm?

Mobile phones store charge in capacitors, made from tantalum metal. Tantalum is extracted from a mineral called coltan. Miners in the Democratic Republic of Congo find coltan in streams. It's worth a lot of money.

5

So that's another benefit.

True. But there are impacts on the environment. And they're not good. Mining companies have cut down forests to make it easier to find and transport coltan. The forests are home to mountain gorillas. What will happen to this rare species if they destroy more forests?

6

But can't scientists be part of the solution? Maybe they can work out how to rebuild the gorilla population. Perhaps they could develop better ways of recycling tantalum from phones.

Maybe. But for now, weigh the costs against the benefits. And don't buy that phone you don't need.

Making decisions about science and technology

1   A company wants to open a new bauxite mine in Jamaica.

Aluminium metal is extracted from bauxite ore.

Different people have different opinions about the mine.

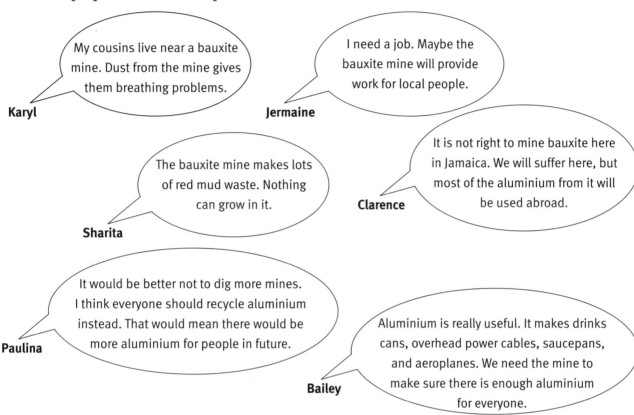

Karyl: My cousins live near a bauxite mine. Dust from the mine gives them breathing problems.

Jermaine: I need a job. Maybe the bauxite mine will provide work for local people.

Sharita: The bauxite mine makes lots of red mud waste. Nothing can grow in it.

Clarence: It is not right to mine bauxite here in Jamaica. We will suffer here, but most of the aluminium from it will be used abroad.

Paulina: It would be better not to dig more mines. I think everyone should recycle aluminium instead. That would mean there would be more aluminium for people in future.

Bailey: Aluminium is really useful. It makes drinks cans, overhead power cables, saucepans, and aeroplanes. We need the mine to make sure there is enough aluminium for everyone.

6

a   Write the names of the people above in the correct box in the table below.

Each name may be used once, more than once, or not at all.

You can write one or more names in each box.

A person or people who identify...	Name or names
...an impact on the environment	
...an issue that science cannot solve	
...issues that could be investigated scientifically	
... an ethical issue	
...an unintended impact on the environment	
...an issue linked to sustainability	

b Some friends are discussing whether it is better to extract aluminium from bauxite ore, or whether it is better to recycle the metal.

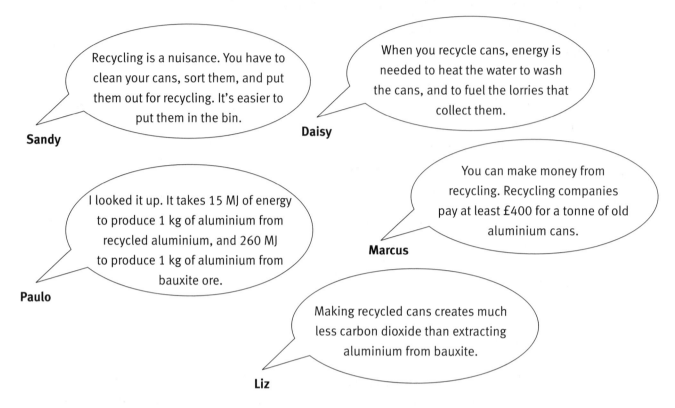

Sandy: Recycling is a nuisance. You have to clean your cans, sort them, and put them out for recycling. It's easier to put them in the bin.

Daisy: When you recycle cans, energy is needed to heat the water to wash the cans, and to fuel the lorries that collect them.

Paulo: I looked it up. It takes 15 MJ of energy to produce 1 kg of aluminium from recycled aluminium, and 260 MJ to produce 1 kg of aluminium from bauxite ore.

Marcus: You can make money from recycling. Recycling companies pay at least £400 for a tonne of old aluminium cans.

Liz: Making recycled cans creates much less carbon dioxide than extracting aluminium from bauxite.

i Use the data and opinions above and on the previous page to help you fill in the grid below.

	Benefits	Drawbacks
Extracting aluminium from bauxite ore		
Recycling aluminium		

ii Use the table to write a paragraph to compare the benefits and drawbacks of extracting aluminium from bauxite ore compared to recycling aluminium.

Identify which method you think is better, and give reasons for your choice.

_____

_____

_____

_____

Making decisions about science and technology

1   Graphite is a mineral. It is made up of carbon atoms.
    A European company wants to reopen an old graphite mine in
    Mozambique, Africa.

    Some people give their opinions about re-opening the mine.
    Use the opinions to help you answer the questions below.

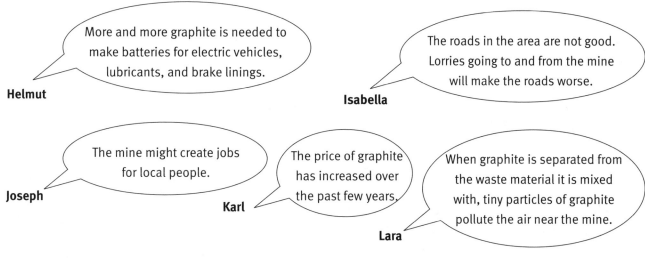

Helmut: More and more graphite is needed to make batteries for electric vehicles, lubricants, and brake linings.

Isabella: The roads in the area are not good. Lorries going to and from the mine will make the roads worse.

Joseph: The mine might create jobs for local people.

Karl: The price of graphite has increased over the past few years.

Lara: When graphite is separated from the waste material it is mixed with, tiny particles of graphite pollute the air near the mine.

   a   Identify two groups of people who might benefit from the mine,
           and suggest one way in which each group might benefit.

   _____

   _____ [2]

   b   Identify two problems that might be caused if the mine re-opens.

   _____

   _____ [2]

   c   People in Mozambique have some questions about the benefits
           and problems of reopening the mine.

           Which of the questions below could be answered using a
           scientific approach?

           Tick the correct boxes.

           What mass of graphite is in 1 tonne of rock from
           the mine?   ☐

           How much money would the mining company make?   ☐

           Is the graphite of good enough quality to make
           batteries?   ☐

           How does graphite dust affect health?   ☐

           Overall, would opening the mine benefit
           local people?   ☐ [2]

6

d Look at the data in the table.

Mass of graphite extracted from the Earth's crust in 2008 (tonnes)	1 million
Estimated total mass of graphite that could be extracted from the Earth's crust (tonnes)	800 million

    i If the same mass of graphite that was extracted in 2008 is extracted every year in future, after how many years will the Earth's graphite run out?

       _____ years   [1]

    ii Graphite can be recycled.
       Explain why it is more sustainable to recycle graphite than to extract graphite from mines.

       _____

       _____   [2]

e In the UK, the law states that people working with graphite must not be exposed to more than 10 mg/m³ of graphite dust on average, over 8 hours.

    i Suggest why there is a law to limit how much graphite dust people can be exposed to.

       _____

       _____   [1]

    ii Suggest what data scientists collected to help the government decide on the legal limit for graphite dust exposure.

       _____

       _____   [1]

Total [11]

abundant Abundance measures how common an element is. Silicon is abundant in the lithosphere. Nitrogen is abundant in the atmosphere.

accumulate To collect together and increase in quantity.

accuracy How close a quantitative result is to the true or 'actual' value.

acid A compound that dissolves in water to give a solution with a pH lower than 7. Acid solutions change the colour of indicators, form salts when they neutralise alkalis, react with carbonates to form carbon dioxide, and give off hydrogen when they react with a metal. An acid is a compound that contains hydrogen in its formula and produces hydrogen ions when it dissolves in water.

activation energy The minimum energy needed in a collision between molecules if they are to react. The activation energy is the height of the energy barrier between reactants and products in a chemical change.

actual yield The mass of the required chemical obtained after separating and purifying the product of a chemical reaction.

alcohols Alcohols are organic compounds containing the reactive group —OH. Ethanol is an alcohol. It has the formula C_2H_5OH.

alkali A compound that dissolves in water to give a solution with a pH higher than 7. An alkali can be neutralised by an acid to form a salt.
Ⓗ Solutions of alkalis contain hydroxide ions.

Alkali Acts Acts of Parliament passed in the UK in order to control levels of pollution. They led to the formation of an Alkali Inspectorate, which checked that at least 95% of acid fumes were removed from the chimneys of chemical factories.

alkali metal An element in Group 1 of the periodic table. Alkali metals react with water to form alkaline solutions of the metal hydroxide.

alkane Alkanes are hydrocarbons found in crude oil. All the C—C bonds in alkanes are single bonds. Ethane is an alkane. It has the formula C_2H_6.

alkene Alkenes are hydrocarbons that contain a C=C double bond. Ethene is an alkene. It has the formula C_2H_4.

alloy A mixture of metals. Alloys are often more useful than pure metals.

aqueous An aqueous solution is a solution in which water is the solvent.

assumption A piece of information that is taken for granted without sufficient evidence to be certain.

atmosphere The layer of gases that surrounds the Earth.

atom The smallest particle of an element. The atoms of each element are the same as each other and are different from the atoms of other elements.

atom economy A measure of the efficiency of a chemical process. The atom economy for a process shows the mass of product atoms as a percentage of the mass of reactant atoms.

attractive forces (between molecules) Forces that try to pull molecules together. Attractions between molecules are weak. Molecular chemicals have low melting points and boiling points because the molecules are easy to separate.

balanced equation An equation showing the formulae of the reactants and products. The equation is balanced when there is the same number of each kind of atom on both sides of the equation.

best estimate When measuring a variable, the value in which you have most confidence.

biodegradable Materials that are broken down in the environment by microorganisms. Most synthetic polymers are not biodegradable.

biomass Plant material and animal waste that can be used as a fuel. A renewable energy source.

bleach A chemical that can destroy unwanted colours. Bleaches also kill bacteria. A common bleach is a solution of chlorine in sodium hydroxide.

bond strength A measure of how much energy is needed to break a covalent bond between two atoms. It is measured in joules.

branched chains Chains of carbon atoms with short side branches.

brine A solution of sodium chloride (salt) in water. Brine is produced by solution mining of underground salt deposits.

bulk chemicals Chemicals made by industry on a scale of thousands or millions of tonnes per year. Examples are sulfuric acid, nitric acid, sodium hydroxide, ethanol, and ethanoic acid.

burette A graduated tube with taps or valves used to measure the volume of liquids or solutions during quantitative investigations such as titrations.

by-products Unwanted products of chemical synthesis. By-products are formed by side-reactions that happen at the same time as the main reaction, thus reducing the yield of the product required.

Ⓗ**calculated risk** Risk calculated from reliable data.

carbonate A compound that contains carbonate ions, $CO_3{}^{2-}$. An example is calcium carbonate, $CaCO_3$.

carboxylic acid Carboxylic acids are organic compounds containing the reactive group —COOH. Ethanoic acid (acetic acid) is an example. It has the formula CH_3COOH.

carrier gas The mobile phase in gas chromatography.

catalyst A chemical that speeds up a chemical reaction but is not used up in the process.

catalytic converter A device fitted to a vehicle exhaust that changes the waste gases into less harmful ones.

cause When there is evidence that changes in a factor produce a particular outcome, then the factor is said to cause the outcome. For example, increases in the pollen count cause increases in the incidence of hay fever.

centrifuge A piece of equipment used to separate a mixture of liquids and solids by spinning the mixture very fast.

ceramic Solid materials such as pottery, glass, cement, and brick.

chemical change/reaction A change that forms a new chemical.

chemical equation A summary of a chemical reaction showing the reactants and products with their physical states (see balanced chemical equation).

chemical formula A way of describing a chemical that uses symbols for atoms. It gives information about the number of different types of atom in the chemical.

Glossary

chemical industry The industry that converts raw materials such as crude oil, natural gas, and minerals into useful products such as pharmaceuticals, fertilisers, paints, and dyes.

chemical properties A chemical property describes how an element or compound interacts with other chemicals, for example, the reactivity of a metal with water.

chemical species The different chemical forms that an element can take. For example, chlorine has three chemical species: atom, molecule, and ion. Each of these forms has distinct properties.

(H) chemical synthesis Making a new chemical by joining together simpler chemicals.

chlorination The process of adding chlorine to water to kill microorganisms, so that it is safe to drink.

chlorine A greenish toxic gas, used to bleach paper and textiles, and to treat water.

chromatogram The resulting record showing the separated chemicals at the end of a chromatography experiment.

chromatography An analytical technique in which the components of a mixture are separated by the movement of a mobile phase through a stationary phase.

collision theory The theory that reactions happen when molecules collide. The theory helps to explain the factors that affect the rates of chemical change. Not all collisions between molecules lead to reaction.

combustion When a chemical reacts rapidly with oxygen, releasing energy.

compression A material is in compression when forces are trying to push it together and make it smaller.

concentration The quantity of a chemical dissolved in a stated volume of solution. Concentrations can be measured in grams per litre.

condensed The change of state from a gas to a liquid, for example, water vapour in the air condenses to form rain.

conservation of atoms All the atoms present at the beginning of a chemical reaction are still there at the end. No new atoms are created and no atoms are destroyed during a chemical reaction.

conservation of mass The total mass of chemicals is the same at the end of a reaction as at the beginning. No atoms are created or destroyed and so no mass is gained or lost.

convection The movement that occurs when hot material rises and cooler material sinks.

correlation When an outcome happens if a specific factor is present, but does not happen when it is absent, or if a measured outcome increases (or decreases steadily) as the value of a factor increases, there is a correlation between the two. For example, a matching pattern in the variation of pollen count and the incidence of hayfever is evidence of a correlation.

corrosive A corrosive chemical may destroy living tissue on contact.

covalent bonding Strong attractive forces that hold atoms together in molecules. Covalent bonds form between atoms of non-metallic elements.

cross-links Links or bonds joining polymer chains together.

crude oil A dark, oily liquid found in the Earth, which is a mixture of hydrocarbons.

crust (of the Earth) The outer layer of the lithosphere.

crystalline A material with molecules, atoms, or ions lined up in a regular way as in a crystal.

crystalline polymer A polymer with molecules lined up in a regular way as in a crystal.

crystallise To form crystals, for example, by evaporating the water from a solution of a salt.

denatured When the shape of an enzyme has been changed, usually as a result of external temperatures being too high or pH changes. The enzyme no longer works.

density A dense material is heavy for its size. Density is mass divided by volume.

diamond A gemstone. A form of carbon. It has a giant covalent structure and is very hard.

diatomic A molecule with two atoms, for example, N_2, O_2, Cl_2.

displacement reaction A more reactive halogen will displace a less reactive halogen, for example, chlorine will displace bromide ions to form bromine and chloride ions.

dissolve Some chemicals dissolve in liquids (solvents). Salt and sugar, for example, dissolve in water.

distillation A method of separating a mixture of two or more substances with different boiling points.

double bond A covalent bond between two atoms involving the sharing of two pairs of electrons. They are found in alkenes and unsaturated hydrocarbons.

drying agent A chemical used to remove water from moist liquids or gases. Anhydrous calcium chloride and anhydrous sodium sulfate are examples of drying agents.

durable A material is durable if it lasts a long time in use. It does not wear out.

(H) dynamic equilibrium Chemical equilibria are dynamic. At equilibrium the forward and back reactions are still continuing but at equal rates so that there is no overall change.

(H) economic context How money changes hands between businesses, government, and individuals.

efficiency The percentage of energy supplied to a machine that is usefully transferred by it.

electrode A conductor made of a metal or graphite through which a current enters or leaves a chemical during electrolysis. Electrons flow into the negative electrode (cathode) and out of the positive electrode (anode).

electrolysis Splitting up a chemical into its elements by passing an electric current through it.

electrolyte A chemical that can be split up by an electric current when molten or in solution is the electrolyte. Ionic compounds are electrolytes.

electron arrangement The number and arrangement of electrons in an atom of an element.

electrons Tiny particles in atoms. Electrons are found outside the nucleus. Electrons have negligible mass and are negatively charged, 1–.

electrostatic attraction The force of attraction between objects with opposite electric charges. A positive ion, for example, attracts a negative ion.

emission Something given out by something else, for example, the emission of carbon dioxide from combustion engines.

end point The point during a titration at which the reaction is just complete. For example, in an acid–alkali titration, the end point is reached when the indicator changes colour. This happens when exactly the right amount of acid has been added to react with all the alkali present at the start.

endothermic An endothermic process takes in energy from its surroundings.

energy level The electrons in an atom have different energies and are arranged at distinct energy levels.

energy-level diagram A diagram to show the difference in energy between the reactants and the products of a reaction.

enzyme A biological catalyst.

equilibrium A state of balance in a reversible reaction when neither the forward nor the backward reaction is complete. The reaction appears to have stopped. At equilibrium reactants and products are present and their concentrations are not changing.

erosion The movement of solids at the Earth's surface (for example, soil, mud, and rock) caused by wind, water, ice, gravity, and living organisms.

esters An organic compound made from a carboxylic acid and an alcohol. Ethyl ethanoate is an ester. It has the formula $CH_3COOC_2H_5$.

ethics A set of principles that may show how to behave in a situation.

evaporate The change of state from a liquid to a gas.

exothermic An exothermic process gives out energy to its surroundings.

extraction (of metals) The process of obtaining a metal from a mineral by chemical reduction or electrolysis. It is often necessary to concentrate the ore before extracting the metal.

extruded A plastic is shaped by being forced through a mould.

factor A variable that changes and may affect something else.

fat Fats are esters of glycerol with long-chain carboxylic acids (fatty acids). The fatty acids in animal fats are mainly saturated compounds.

fatty acids Another name for carboxylic acids.

feedstocks A chemical, or mixture of chemicals, fed into a process in the chemical industry.

fermentation The conversion of carbohydrates to alcohols and carbon dioxide using yeast.

fibres Long thin threads that make up materials such as wool and polyester. Most fibres used for textiles consist of natural or synthetic polymers.

filter To separate a solid from a liquid by passing it through a filter paper.

fine chemicals Fine chemicals are used in products such as food additives, medicines, and pesticides.

flame colour A colour produced when a chemical is held in a flame. Some elements and their compounds give characteristic colours. Sodium and sodium compounds, for example, give bright-yellow flames.

flavouring Mixtures of chemicals that give food, sweets, toothpaste, and other products their flavours.

flexible A flexible material bends easily without breaking.

formula (chemical) A way of describing a chemical that uses symbols for atoms. A formula gives information about the numbers of different types of atom in the chemical. The formula of sulfuric acid, for example, is H_2SO_4.

fossils The stony remains of an animal or plant that lived millions of years ago, or an imprint it has made (for example, a footprint) in a surface.

fractional distillation The process of separating crude oil into groups of molecules with similar boiling points called fractions.

fractions A mixture of hydrocarbons with similar boiling points that have been separated from crude oil by fractional distillation.

functional group A reactive group of atoms in an organic molecule. The hydrocarbon chain making up the rest of the molecule is generally unreactive with common reagents such as acids and alkalis. Examples of functional groups are —OH in alcohols and —COOH in carboxylic acids.

giant covalent structure A giant, three-dimensional arrangement of atoms that are held together by covalent bonds. Silicon dioxide and diamond have giant covalent structures.

giant ionic structure The structure of solid ionic compounds. There are no individual molecules, but millions of oppositely charged ions packed closely together in a regular, three-dimensional arrangement.

glycerol Glycerol is an alcohol with three —OH groups. Its chemical name is propan-1,2,3-triol. Its formula is CH_2OH—$CHOH$—CH_2OH.

grain Relatively small particle of a substance, for example, grains of sand.

graphite A form of carbon. It has a giant covalent structure. It is unusual for a non-metal in that it conducts electricity.

group Each column in the periodic table is a group of similar elements.

Haber process The reaction between nitrogen and hydrogen gas used to make ammonia on an industrial scale.

halogens The family name of the Group 7 elements.

hard A material that is difficult to dent or scratch.

harmful A harmful chemical is one that may cause damage to health if swallowed, breathed in, or absorbed through the skin.

heat under reflux Heating a reaction mixture in a flask fitted with a vertical condenser. Vapours escaping from the flask condense and flow back into the reaction mixture.

hydrocarbon A compound of hydrogen and carbon only. Ethane, C_2H_6, is a hydrocarbon.

hydrogen chloride gas An acid gas that is toxic and corrosive, and is produced by the Leblanc process.

hydrogen ion A hydrogen atom that has lost one electron. The symbol for a hydrogen ion is H⁺. Acids produce aqueous hydrogen ions, H⁺(aq), when dissolved in water.

hydrogen sulfide gas A poisonous gas that smells of rotten eggs.

hydrosphere All the water on Earth. This includes oceans, lakes, rivers, underground reservoirs, and rainwater.

hydroxide ion A negative ion, OH⁻. Alkalis give aqueous hydroxide ions when they dissolve in water.

incinerator A factory for burning rubbish and generating electricity.

indicator A chemical that shows whether a solution is acidic or alkaline. For example, litmus turns blue in alkalis and red in acids. Universal indicator has a range of colours that show the pH of a solution.

ion An electrically charged atom or group of atoms.

ionic bonding Very strong attractive forces that hold the ions together in an ionic compound. The forces come from the attraction between positively and negatively charged ions.

ionic compounds Compounds formed by the combination of a metal and a non-metal. They contain positively charged metal ions and negatively charged non-metal ions.

ionic equation An ionic equation describes a chemical change by showing only the reacting ions in solution, for example, $Ba^{2+}(aq) + SO_4^{2-}(aq) \longrightarrow BaSO_4(s)$

irreversible change A chemical change that can only go in one direction, for example, changes involving combustion.

landfill Disposing of rubbish in holes in the ground.

latitude The location of a place on Earth, north or south of the equator.

leach The movement of the plasticisers in a polymer into water, or another liquid, that is flowing past the polymer or is contained by it.

Leblanc process A process that used chalk (calcium carbonate), salt (sodium chloride) and coal to make the alkali, sodium carbonate. The Leblanc process was highly polluting.

Le Chatelier's principle The principle that the position of an equilibrium will respond to oppose a change in the reaction conditions.

life cycle assessment A way of analysing the production, use, and disposal of a material or product to add up the total energy and water used and the effects on the environment.

limiting factor The factor that prevents the rate of growth of living things.

line spectrum A spectrum made up of a series of lines. Each element has its own characteristic line spectrum.

lithosphere The rigid outer layer of the Earth, made up of the crust and the part of mantle just below it.

locating agent A chemical used to show up colourless spots on a chromatogram.

long-chain molecule Polymers are long-chain molecules. They consist of long chains of atoms.

macroscopic Large enough to be seen without the help of a microscope.

magnetic A material that is attracted to a magnet. For example, iron is magnetic.

mantle The layer of rock between the crust and the outer core of the Earth. It is approximately 2900 km thick.

material The polymers, metals, glasses, and ceramics that we use to make all sorts of objects and structures.

mean value A type of average, found by adding up a set of measurements and then dividing by the number of measurements. You can have more confidence in the mean of a set of measurements than in a single measurement.

measurement uncertainty Variations in analytical results owing to factors that the analyst cannot control. Measurement uncertainty arises from both systematic and random errors.

melting point The temperature at which something melts.

metal An element on the left side of the periodic table. Metals have characteristic properties: they are shiny when polished and they conduct electricity. Some metals react with acids to give salts and hydrogen. Metals are present as positive ions in salts.

metal hydroxide A compound consisting of metal positive ions and hydroxide ions. Examples are sodium hydroxide, NaOH, and magnesium hydroxide, $Mg(OH)_2$.

metal oxide A compound of a metal with oxygen.

metallic bonding Very strong attractive forces that hold metal atoms together in a solid metal.
Ⓗ The metal atoms lose their outer electrons and form positive ions. The electrons drift freely around the lattice of positive metal ions and hold the ions together.

microorganisms Living organisms that can only be seen by looking at them through a microscope. They include bacteria, viruses, and fungi.

mineral A naturally occurring element or compound in the Earth's lithosphere.

mixture Two or more different chemicals, mixed but not chemically joined together.

mobile phase The solvent that carries chemicals from a sample through a chromatographic column or sheet.

molecular model Model to show the arrangement of atoms in a molecule, and the bonds between the atoms.

molecule A group of atoms joined together. Most non-metals consist of molecules. Most compounds of non-metals with other non-metals are also molecular.

molten A chemical in the liquid state. A chemical is molten when the temperature is above its melting point but below its boiling point.

monomer A small molecule that can be joined to others like it in long chains to make a polymer.

nanometre A unit of length 1000 000 000 times smaller than a metre.

nanoparticle A very tiny particle, whose size can be measured in nanometres.

nanotechnology The use and control of matter on a tiny (nanometre) scale.

natural A material that occurs naturally but may need processing to make it useful, such as silk, cotton, leather, and asbestos.

negative ion An ion that has a negative charge (an anion).

neutralisation A reaction in which an acid reacts with an alkali to form a salt.
During neutralisation reactions, the hydrogen ions in the acid solution react with hydroxide ions in the alkaline solution to make water molecules.

neutrons An uncharged particle found in the nucleus of atoms. The relative mass of a neutron is 1.

nitrogen cycle The continual cycling of nitrogen, which is one of the elements that is essential for life. By being converted to different chemical forms, nitrogen is able to cycle between the atmosphere, lithosphere, hydrosphere, and biosphere.

nitrogen fixation The conversion of nitrogen gas into compounds either industrially or by natural means.

nitrogenase The enzyme system that catalyses the reduction of nitrogen gas to ammonia.

non-aqueous A solution in which a liquid other than water is the solvent.

nucleus The tiny central part of an atom (made up of protons and neutrons). Most of the mass of an atom is concentrated in its nucleus.

ore A natural mineral that contains enough valuable minerals to make it profitable to mine.

organic chemistry The study of carbon compounds. This includes all of the natural carbon compounds from living things and synthetic carbon compounds.

organic matter Material that has come from dead plants and animals.

outcome A variable that changes as a result of something else changing.

outlier A measured result that seems very different from other repeat measurements, or from the value you would expect, which you therefore strongly suspect is wrong.

oxidation A reaction that adds oxygen to a chemical.

oxide A compound of an element with oxygen.

particulate Tiny bit of a solid.

percentage yield A measure of the efficiency of a chemical synthesis.

perceived risk The level of risk that people think is attached to an activity, not based on data.

period In the context of chemistry, a row in the periodic table.

periodic In chemistry, a repeating pattern in the properties of elements. In the periodic table one pattern is that each period starts with metals on the left and ends with non-metals on the right.

persistent organic pollutants (POPs) POPs are organic compounds that do not break down in the environment for a very long time. They can spread widely around the world and build up in the fatty tissue of humans and animals. They can be harmful to people and the environment.

petrochemical Chemicals made from crude oil (petroleum) or natural gas.

photosynthesis A chemical reaction that happens in green plants using the energy in sunlight. The plant takes in water and carbon dioxide, and uses sunlight to convert them to glucose (a nutrient) and oxygen.

pH scale A number scale that shows the acidity or alkalinity of a solution in water.

phthalate A chemical that is used as a plasticiser, added to polymers to make them more flexible.

physical properties Properties of elements and compounds such as melting point, density, and electrical conductivity. These are properties that do not involve one chemical turning into another.

pipette A pipette is used to measure small volumes of liquids or solutions accurately. A pipette can be used to deliver the same fixed volume of solution again and again during a series of titrations.

pilot plant A small-scale chemical processing facility. A pilot plant is used to test processes before scaling up to full-scale production.

plant A chemical plant is an industrial facility used to manufacture chemicals.

plasticiser A chemical (usually a small molecule) added to a polymer to make it more flexible.

pollutant Waste matter that contaminates the water, air, or soil.

polymer A material made of very long molecules formed by joining lots of small molecules, called monomers, together.

polymerise The joining together of lots of small molecules called monomers to form a long-chain molecule called a polymer.

positive ions Ions that have a positive charge (cations).

precipitate An insoluble solid formed on mixing two solutions. Silver bromide forms as a precipitate on mixing solutions of silver nitrate and potassium bromide.

precision A measure of the spread of quantitative results. If the measurements are precise all the results are very close in value.

preservative A chemical added to food to stop it going bad.

product The new chemicals formed during a chemical reaction.

properties Physical or chemical characteristics of a chemical. The properties of a chemical are what make it different from other chemicals.

proportional Two variables are proportional if there is a constant ratio between them.

proton number The number of protons in the nucleus of an atom (also called the atomic number). In an uncharged atom this also gives the number of electrons.

protons Tiny particles that are present in the nuclei of atoms. Protons are positively charged, 1+.

qualitative Qualitative analysis is any method for identifying the chemicals in a sample. Thin-layer chromatography is an example of a qualitative method of analysis.

Glossary

quantitative Quantitative analysis is any method for determining the amount of a chemical in a sample. An acid–base titration is an example of quantitative analysis.

range The difference between the highest and the lowest of a set of measurements.

rate of reaction A measure of how quickly a reaction happens. Rates can be measured by following the disappearance of a reactant or the formation of a product.

reactant The chemicals on the left-hand side of an equation. These chemicals react to form the products.

reacting mass The masses of chemicals that react together, and the masses of products that are formed. Reacting masses are calculated from the balanced symbol equation using relative atomic masses and relative formula masses.

reactive metal A metal with a strong tendency to react with chemicals such as oxygen, water, and acids. The more reactive a metal, the more strongly it joins with other elements such as oxygen. So reactive metals are hard to extract from their ores.

Ⓗ real difference The difference between two mean values is real if their ranges do not overlap.

recycling A range of methods for making new materials from materials that have already been used.

reducing agent A chemical that removes oxygen from another chemical. For example, carbon acts as a reducing agent when it removes oxygen from a metal oxide. The carbon is oxidised to carbon monoxide during this process.

reduction A reaction that removes oxygen from a chemical.

reference materials Known chemicals used in analysis for comparison with unknown chemicals.

regulations Rules that can be enforced by an authority, for example, the government. The law that says that all vehicles that are three years old and older must have an annual exhaust emission test is a regulation that helps to reduce atmospheric pollution.

relative atomic mass The mass of an atom of an element compared to the mass of an atom of carbon. The relative atomic mass of carbon is defined as 12.

relative formula mass The combined relative atomic masses of all the atoms in a formula. To find the relative formula mass of a chemical, you just add up the relative atomic masses of the atoms in the formula.

renewable resource Resources that can be replaced as quickly as they are used. An example is wood from the growth of trees.

repeatable A quality of a measurement that gives the same result when repeated under the same conditions.

replicate sample Two or more samples taken from the same material. Replicate samples should be as similar as possible and analysed by the same procedure to help judge the precision of the analysis.

representative sample A sample of a material that is as nearly identical as possible in its chemical composition to that of the larger bulk of material sampled.

Ⓗ reproducible A quality of a measurement that gives the same result when carried out under different conditions, for example, by different people or using different equipment or methods.

retardation factor A retardation factor, R_f, is a ratio used in paper or thin-layer chromatography. If the conditions are kept the same, each chemical in a mixture will move a fixed fraction of the distance moved by the solvent front. The R_f value is a measure of this fraction.

retention time In chromatography, the time it takes for a component in a mixture to pass through the stationary phase.

risk The chance that a hazardous substance or process will harm someone.

risk assessment A check on the hazards involved in a scientific procedure. A full assessment includes the steps to be taken to avoid or reduce the risks from the hazards identified.

rock A naturally occurring solid, made up of one or more minerals.

rubber A material that is easily stretched or bent. Natural rubber is a natural polymer obtained from latex, the sap of a rubber tree.

salt An ionic compound formed when an acid neutralises an alkali or when a metal reacts with a non-metal.

sample A small portion collected from a larger bulk of material for laboratory analysis (such as a water sample or a soil sample).

saturated In the molecules of a saturated compound, all of the bonds are single bonds. The fatty acids in animal fats are all saturated compounds.

scale up To redesign a synthesis to produce a chemical in larger amounts. A process might be scaled up first from a laboratory method to a pilot plant, then from a pilot plant to a full-scale industrial process.

sedimentary rock Rock formed from layers of sediment.

shell A region in space (around the nucleus of an atom) where there can be electrons.

small molecules Particles of chemicals that consist of small numbers of atoms bonded together. Chemicals made up of one or more non-metallic elements and that have low boiling and melting points consist of small molecules.

social context The situation of people's lives.

soft A material that is easy to dent or scratch.

solution Formed when a solid, liquid, or gas dissolves in a solvent.

solvent front The furthest position reached by the solvent during paper or thin-layer chromatography.

spectroscopy The use of instruments to produce and analyse spectra. Chemists use spectroscopy to study the composition, structure, and bonding of elements and compounds.

standard solution A solution whose concentration is accurately known. They are used in titrations.

stationary phase The medium through which the mobile phase passes in chromatography.

stiff A material that is difficult to bend or stretch.

strong A material that is hard to pull apart or crush.

strong acid A strong acid is fully ionised to produce hydrogen ions when it dissolves in water.

subatomic particle The particles that make up atoms. Protons, neutrons, and electrons are subatomic particles.

subsidence The sinking of the ground's surface when it collapses into a hole beneath it.

surface area How much exposed surface a solid object has.

surface area (of a solid chemical) The area of a solid in contact with other reactants that are liquids or gases.

sustainability Using resources and the environment to meet the needs of people today without damaging Earth or reducing the resources for people in future.

sustainable Meeting the needs of today without damaging the Earth for future generations.

sustainable development A plan for meeting people's present needs without spoiling the environment for the future.

synthetic A material made by a chemical process, not naturally occurring.

tap funnel A funnel with a tap to allow the controlled release of a liquid.

tarnish When the surface of a metal becomes dull or discoloured because it has reacted with the oxygen in the air.

tectonic plates Giant slabs of rock (about 12, comprising crust and upper mantle) that make up the Earth's outer layer.

tension A material is in tension when forces are trying to stretch it or pull it apart.

theoretical yield The amount of product that would be obtained in a reaction if all the reactants were converted to products exactly as described by the balanced chemical equation.

theory A scientific explanation that is generally accepted by the scientific community.

titration An analytical technique used to find the exact volumes of solutions that react with each other.

toxic A chemical that may lead to serious health risks, or even death, if breathed in, swallowed, or taken in through the skin.

toxin A poisonous chemical produced by a microorganism, plant, or animal.

trend A description of the way a property increases or decreases along a series of elements or compounds, which is often applied to the elements (or their compounds) in a group or period.

triple bond A covalent bond between the two atoms involving the sharing of three pairs of electrons, for example, nitrogen gas. It makes the molecule very stable and unreactive.

uncertain Describes measurements where scientists know that they may not have recorded the true value.

uncertainty The amount by which a measurement could differ from the true value.

unsaturated There are double bonds in the molecules of unsaturated compounds. There is no spare bonding. The fatty acids in vegetable oils include a high proportion of unsaturated compounds.

vegetable oil Vegetable oils are esters of glycerol with fatty acids (long-chain carboxylic acids). More of the fatty acids in vegetable oils are unsaturated when compared with the fatty acids in animal fats.

vinegar A sour-tasting liquid used as a flavouring and to preserve foods. It is a dilute acetic (ethanoic) acid made by fermenting beer, wine, or cider.

vulcanisation A process for hardening natural rubber by making cross-links between the polymer molecules.

weak acids Weak acids are only slightly ionised to produce hydrogen ions when they dissolve in water.

H **wet scrubbing** A process used to remove pollutants from flue gases.

word equation A summary in words of a chemical reaction.

C1 Workout

1. 78% nitrogen; 21% oxygen; 1% argon
2. The early atmosphere was mainly **carbon dioxide** and **water vapour**.
   Water vapour **condensed** to form oceans. Carbon dioxide **dissolved** in the oceans. Later it formed **sedimentary** rocks. Early plants removed **carbon dioxide** from the atmosphere by photosynthesis, and added **oxygen** to the atmosphere.
3. Car A, going into engine: N_2, O_2; coming out of exhaust: CO_2, H_2O, N_2
   Car B, going into engine: N_2, O_2; coming out of exhaust: CO_2, H_2O, N_2, NO, NO_2, C, CO
4. Carbon or hydrogen; hydrogen or carbon; oxygen; carbon dioxide; water; chemical or combustion or burning; number; products; rearranged; products; products

5.

	Reactants		Products
Name	coal (with no sulfur impurities)	oxygen (from a plentiful supply of air)	carbon dioxide
Formula	C	O_2	CO_2
Diagram			

	Reactants		Product		
Name	coal (with no sulfur impurities)	oxygen (from a limited supply of air)	carbon dioxide	carbon monoxide	particulate carbon
Formula	C	O_2	CO_2	CO	C
Diagram					

	Reactants		Products	
Name	coal (with sulfur impurities)	oxygen (from a plentiful supply of air)	carbon dioxide	sulfur dioxide
Formula	C and S	O_2	CO_2	SO_2
Diagram				

6. Suggested answers include:
   We should replace diesel and petrol with biofuels because the plants from which biofuels are made remove carbon dioxide from the atmosphere.
   No we shouldn't. The problem with biofuels is that they produce carbon dioxide when they burn.
   I think electric vehicles are better because they produce no pollutants as they travel.
   True, but there are problems with electric vehicles. One is that the electricity may have been generated by burning fossil fuels.

7.

Pollutant name	Pollutant formula	Where the pollutant comes from	Problems the pollutant causes	One way of reducing the amount of this pollutant added to the atmosphere
sulfur dioxide	SO_2	burning fossil fuels with sulfur impurities	acid rain	remove sulfur impurities from fuels before burning
nitrogen oxides	NO_2 NO	burning fuels in car engines	acid rain asthma	fit catalytic converters to cars
carbon dioxide	CO_2	burning fossil fuels	global warming	burn less fossil fuels
carbon monoxide	CO	burning fossil fuels in a limited supply of oxygen	poisoning	make sure oxygen supply is plentiful
particulate carbon	C	burning fossil fuels in a limited supply of oxygen	makes surfaces dirty	make sure oxygen supply is plentiful

C1 Quickfire

1. Sulfur dioxide – acid rain
   Carbon dioxide – climate change
   Carbon monoxide – reduces the amount of oxygen the blood carries
   Particulate carbon – makes surfaces dirty
2. Argon – 1%; nitrogen – 78%; oxygen – 21%
3. Carbon, hydrocarbons, hydrogen, oxygen, carbon dioxide
4. True: **a**, **d**, **e**
   Corrected versions of false sentences:
   b The spaces between molecules in the air are large.
   c Carbon monoxide is directly harmful to humans.
5. C, B, D, A
6. Carbon dioxide dissolved in the oceans, and some of its carbon atoms ended up in sedimentary rocks. Early plants used carbon dioxide for photosynthesis.
7. Hydrocarbon, oxygen, oxidation, reduction

8.

Formula	Diagram of molecule	Name
CO		carbon monoxide
SO_2		sulfur dioxide
CO_2		carbon dioxide
NO_2		nitrogen dioxide
H_2O		water
NO		nitrogen monoxide

9 Have more efficient engines that burn less fuel; use low-sulfur fuels; use catalytic converters; use more public transport; have and enforce legal limits on exhaust emissions

10 The same total number of atoms of each element are present in both the reactants and products; the atoms are rearranged in the reaction.

11 Nitrogen, nitrogen monoxide, NO, nitrogen dioxide, NO_2, NO_x, acid rain

12 Spray seawater at the flue gases. Substances in the seawater react with sulfur dioxide. Allow the flue gases to be in contact with a slurry of calcium sulfate and water. Sulfur dioxide and calcium sulfate react to make calcium sulfate.

13 Benefits: the plants from which biofuels are made remove carbon dioxide from the atmosphere; renewable. Problems: add carbon dioxide to the atmosphere on burning; may be grown on land that could be used for food crops.

14 Electric cars produce no air pollutants at point of use, unlike diesel cars. However, the electricity might have been generated by burning fossil fuels. Electric cars need to be recharged more frequently than diesel cars need to be refuelled.

15 11 g

16 32 g

17 2 g

18 9 g

C1 GCSE-style questions

1 a

	Run 1	Run 2	Run 3	Run 4
Volume of air in syringes at start (cm³)	100	100	100	100
Volume of air in syringes at end (cm³)	82	86	**85**	84
Decrease in volume (cm³)	18	**14**	15	16

b 15.75 cm³

c The percentage of oxygen in the air is 21%, so the volume of oxygen in 100 cm³ air is 21 cm³. You would expect the copper to react with all the oxygen in the syringes.

d Some of the oxygen in the syringes did not react with the copper **or** there was not enough copper to react with all the oxygen from the air.

2 a i

   ii Hydrocarbon fuels with sulfur impurities

   b i Increased

   ii More coal-fired power stations or more vehicles burning hydrocarbon fuels or any other sensible answer

   c i Nitrogen; oxygen

   ii Acid rain damages buildings made of limestone; acid rain makes lakes more acidic; acid rain damages trees.

   d Sulfur dioxide gas emissions decreased between 1980 and 2000.

3 5/6 marks

Answer clearly describes the correlation shown by the data in the table **and** gives examples to illustrate the trend **and** points out one or more exceptions to the overall trend, giving an example **and** states and explains why it is not possible to say whether increased mass causes increased emissions **All** information in the answer is relevant, clear, organised and presented in a structured and coherent format. Specialist terms are used appropriately. Few, if any, errors in grammar, punctuation and spelling.

3/4 marks

Answer describes the correlation shown by the data in the table, but does not give examples to illustrate the trend **and** points out one or more exceptions to the overall trend **or** states that it is not possible to say whether increased mass causes increased emissions

Most of the information is relevant and presented in a structured and coherent format. Specialist terms are usually used correctly. There are occasional errors in grammar, punctuation and spelling.

1/2 marks

Answer describes the correlation shown by the data in the table, but does not give examples to illustrate the trend **and** does not point out exceptions to the overall trend **or** states that it is not possible to say whether increased mass causes increased emissions

There may be limited use of specialist terms. Errors of grammar, punctuation and spelling prevent communication of the science. Answer includes 1 or 2 points of those listed below.

0 marks

Insufficient or irrelevant science. Answer not worthy of credit.

Relevant points include:

- Overall, there is a correlation between the car mass and average CO_2 emissions – the heavier the car, the greater the mass of average CO_2 emissions...
- ... for example the Fox has a mass of 978 kg and emissions of 144 g/km, and the car with the greatest mass (the Toureg, 2214 kg) has emissions of 324 g/km.
- Not all the data fits the pattern exactly...
- ...for example the Fox has a mass of 978 kg and emissions of 144 g/km, and the Polo has a mass that is greater than the Fox (1000 kg) but smaller emissions of 138 g/km.
- It is not possible to say whether the increased mass causes increased emissions...
- ...because the increased emissions could be caused by some other factor.

C2 Workout

1 Example answers are given here – many others are possible.

Part of tricycle	Properties this part of the tricycle must have	Material
tyres	high frictional forces with ground	rubber
brake	high frictional force with wheel	plastic
frame	high strength	steel
seat	high strength, not cold to the touch	polypropene
handle to push tricycle	high strength, not cold to the touch	polypropene
pushing pole	high strength	aluminium
screws that join pushing pole to handle	resistant to corrosion	stainless steel
bag	flexible	polythene

Answers

2

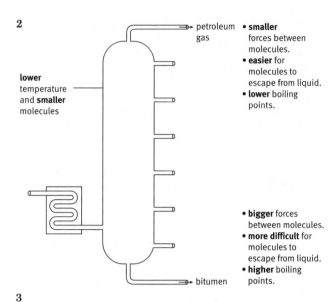

lower temperature and **smaller** molecules

petroleum gas
- **smaller** forces between molecules.
- **easier** for molecules to escape from liquid.
- **lower** boiling points.

- **bigger** forces between molecules.
- **more difficult** for molecules to escape from liquid.
- **higher** boiling points.

bitumen

3

Observation	Suggested reason
Nano-sized particles get into the nuclei of cancer cells more easily than normal-sized gold particles.	Gold nano-particles are much smaller than normal gold particles.

Action	Suggested reason
The scientists tried to stop cancer cells dividing.	They knew this would kill the cancer cells.
The scientists did the tests on cancer cells outside the body.	They did not want to risk harming the patients. It might not have been considered ethical to do the tests on people.
The scientists took cancer cells from many people.	To check that the gold nanoparticles killed cancerous cells from many people, not just from one person.
In future, when the scientists do tests on cancer cells inside the body, they will try to prevent gold nanoparticles entering the nuclei healthy cells.	If gold enters the nucleus of a cell, it is likely to kill the cell, whether or not it is cancerous.

4 Top diagram – make cross-links between polymer chains – harder, stronger, less flexible; second from top diagram – pack molecules neatly together with crystalline regions – stronger, denser; second from bottom diagram – increase chain length – stronger; bottom diagram – add plasticizer – softer, more flexible.

5 a C   b P   c C   d P

6 a One more oxygen molecule
   b Two more carbon dioxide molecules
   c Three more water molecules

C2 Quickfire

1 a, d, e, f
2 True statement: **c**
Corrected versions of false statements:
   a A synthetic material is one which is made from non-living materials.
   b A hydrocarbon is a compound made of carbon and hydrogen only.
   d Monomers are small molecules that join together to form polymers.
   e Most crude oil is used to make fuels.
   f In a chemical reaction, there are always the same number of atoms of each element in the products and in the reactants.

3 Molecules, 1, 100, nm, fuel combustion products, seaspray.
4 The dental polymer matches the colour of teeth, it has no know health risks and is a poor conductor of heat, meaning it does not cause pain when very cold or very hot food and drink are consumed.
5 Polypropene is stronger under tension and does not rot.
6 HDPE is stronger and stiffer.
7 Fractions, greater, higher, lower down.
8 Nanoparticles have a much larger surface area compared to their volume.
9 Silver nanoparticles are added to fibres (in wound dressings and socks, for example) to give antibacterial properties. Nanoparticles are also added to plastics for sports equipment to make them stronger.
10 Nanoparticles may have harmful effects on health. Some people think they should not be widely used until these effects have been fully investigated.
11 a Increased chain length increases strength since longer molecules become more tangled and so more difficult to separate.
   b Cross-linking increases hardness and strength, and decreases flexibility. This is because cross-linking holds the polymer chains together in a rigid pattern.
   c Adding a plasticiser makes a polymer softer and more flexible. This is because the molecules of plasticiser hold the polymer chains apart.
   d Increasing crystallinity increases strength and density because the forces between the molecules are slightly stronger, so more energy is needed to separate them.

C2 GCSE-style questions

1 a Artificial heart valves D, hospital laundry bags C, fillings for front teeth B, contact lenses A
   b i Cellulose
     ii Non-toxic, flexible, high strength in tension
     iii Polymerisation
2 a Density values: aluminium alloy = 2.65 g/cm³; ABS steel = 7.10 g/cm³; glass reinforced plastic = 1.40 g/cm³
   b Points that may be included:
     • Glass reinforced plastic (GRP) and the aluminium alloy have lower densities than ABS steel, so boats of a given size made from GRP and the aluminium alloy will be lighter than a boat of the same size made of ABS steel
     • A boat made from GRP is much more likely to shatter on impact with rocks than one made from the aluminium alloy or the ABS steel
     • The aluminium alloy and ABS steel are stronger than GRP
     • A reasoned choice of material
3 5/6 marks
Answer clearly identifies all the advantages **and** disadvantages of the three materials in the table as building materials, referring to data in the table, **and** comes to a judgement about which is the best material from which to build a house **and** gives clear reasons for the judgement.
All information in the answer is relevant, clear, organised, and presented in a structured and coherent format.
Specialist terms are used appropriately. Few, if any, errors in grammar, punctuation, and spelling.
3/4 marks
Answer identifies some advantages **and** disadvantages of the three materials in the table as building materials, in some cases referring to data in the table
and gives to a judgement about which is the best material from which to build a house.

Most of the information is relevant and presented in a structured and coherent format. Specialist terms are usually used correctly. There are occasional errors in grammar, punctuation and spelling.

1/2 marks

Answer identifies a few advantages **or** disadvantages of some of the materials in the table as building materials, but does not refer to data in the table

or identifies a few advantages **and** disadvantages of one of the materials in the table as a building material, but does not refer to data in the table

or gives a judgement about which material is the best material from which to build a house.

There may be limited use of specialist terms. Errors of grammar, punctuation and spelling prevent communication of the science. Answer includes 1 or 2 points of those listed below.

0 marks

Insufficient or irrelevant science. Answer not worthy of credit.

Relevant points include:

- The compressive strength values for limestone and concrete are the same, at 60 MPa.
- The compressive strength of wood (15 MPa) is much less than that of limestone and concrete (60 MPa)
- Considering strength alone, limestone and concrete are better building materials than wood
- The thermal conductivity of wood (0.1 W/mK) is much less than that of limestone (1.3 W/mK) and concrete (1.7 W/mK)...
- ...This means that wood is a better insulator, and so a house made from wood would lose heat less quickly than an identical house made from limestone or concrete
- A reasoned decision about which material is best, taking into account all the factors above

4 a The forces between long hydrocarbon molecules are stronger than the forces between short hydrocarbon molecules; the stronger the forces between molecules, the more energy is needed to separate them.

   b i Shower curtains, because they need to be most flexible

      ii The rubber in car tyres, because it needs to be more rigid, harder and stronger.

C3 Workout

1

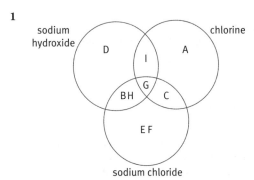

sodium hydroxide

chlorine

D   I   A

G

B H   C

E F

sodium chloride

2 Fossils – different animals lived at different times, so fossils tell us about the ages of the rocks they are in.

Sand grains – comparing sand grains in deserts and rivers to sand grains in sandstones tells us what sort of sand formed the sandstone.

Ripples – the shapes of ripples in rocks give clues about whether the sand was made from river bed sand or desert sand. Shell fragments in limestone tell us about the conditions when the rock formed.

3

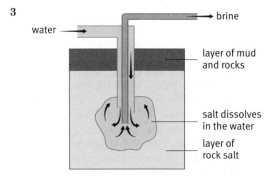

water →

→ brine

layer of mud and rocks

salt dissolves in the water

layer of rock salt

4 1 soap; 2 oxidised; 3 hydroxide; 4 limestone; 5 solution; 6 magnetic; 7 tectonic; 8 hydrogen; 9 flavouring; 10 polymer; 11 chlorine; 12 fossils; 13 distances; 14 pressure

C3 Quickfire

1 a R   b P   c P
   d R   e R

2 Chlorine – to make hydrochloric acid and to make bleach
   Sodium hydroxide – to make soap and to make bleach
   Hydrogen – as a fuel and to make hydrochloric acid
   Sodium chloride – to de-ice roads, to preserve food

3 Polymer, long, carbon, window frames, wire insulation, plasticiser; small, toys

4 Materials, product, energy, chemicals, discarded, energy

5 Sedimentation – with compression, formed limestone when dead sea creatures sank
   Erosion – formed sand, which deposited in layers to form sandstone
   Evaporation – formed rock salt when a sea moved inland
   Mountain building – pushed coal nearer the surface

6 a Tuberculosis, pneumonia (but very similar in both years), typhoid
   b No; some other factor could have caused the decrease in number of deaths from these diseases.

7 They do not break down in the environment; they move long distances in the air and water; they build up in fatty tissues of animals and humans.

8 a Sodium nitrate and water
   b Potassium sulfate and water
   c Potassium chloride, carbon dioxide, and water
   d sodium sulfate, carbon dioxide, and water
   e Potassium chloride and water
   f Sodium nitrate, carbon dioxide, and water

9 a Sodium hydroxide + nitric acid ⟶ sodium nitrate + water
   b Potassium hydroxide + sulfuric acid ⟶ potassium sulfate + water
   c Potassium carbonate + hydrochloric acid ⟶ potassium chloride + carbon dioxide + water
   d Sodium carbonate + sulfuric acid ⟶ sodium sulfate + carbon dioxide + water
   e Potassium hydroxide + hydrochloric acid ⟶ potassium chloride + water
   f Sodium carbonate + nitric acid ⟶ sodium nitrate + carbon dioxide + water

C3 GCSE-style questions

1 a Putting together the computer components in its plastic case – manufacturers make the computer
      Recycling the computer components – people throw away the computer
      Extracting oil from wells beneath the sea – materials are produced
      Dismantling the computer – people throw away the computer
      Making plastics from oil – materials are produced

Answers

b i 1285 kg

ii 5045 kg

iii Two points from: Making the computer requires inputs of materials and energy, and results in the emissions of greenhouse gases. The fewer new computers that are made in a given year, the smaller these inputs and emissions. When a computer is discarded, it is dismantled and the components recycled. These processes use energy. The fewer computers that are discarded in a given year, the smaller the energy input for these processes.

c i Benefit to the environment: extracting metals from rocks leads to larger amounts of waste than using recycled precious metals. Benefit to people: companies that recycle computers can make money by selling their precious metals. *Many other answers are acceptable here.*

ii Risk: inhaling beryllium dust, which is toxic and which may cause cancer. People still dismantle computers because they can sell the valuable metals they contain.

2 5/6 marks

Answer clearly describes each method of manufacturing alkalis

and clearly describes the advantages and disadvantages of each method.

All information in the answer is relevant, clear, organised and presented in a structured and coherent format. Specialist terms are used appropriately. Few, if any, errors in grammar, punctuation and spelling.

3/4 marks

Answer describes each method of manufacturing alkalis

and gives an incomplete account of the advantages and disadvantages of some of the methods

or gives a complete account of the advantages and disadvantages of the methods, but this lacks detail/clarity. Most of the information is relevant and presented in a structured and coherent format. Specialist terms are usually used correctly. There are occasional errors in grammar, punctuation and spelling.

1/2 marks

Answer describes one or two methods of manufacturing alkalis

and gives an advantage and/or disadvantage of one of these methods.

and the explanation lacks detail/clarity.

There may be limited use of specialist terms. Errors of grammar, punctuation and spelling prevent communication of the science. Answer includes 1 or 2 points of those listed below.

0 marks

Insufficient or irrelevant science. Answer not worthy of credit.

Relevant points include:

- alkalis were first made from burnt wood and stale urine...
- ...advantages – materials readily available
- ...disadvantages – the materials were not available in large enough quantities by the time of the industrial revolution
- alkalis were then manufactured from sodium chloride, coal, and limestone...
- ...advantages – could be manufactured in larger quantities
- ...disadvantages – produced pollutants, including hydrogen chloride gas and hydrogen sulfide gas from solid waste produced in the process

- electrolysis of brine...
- ...advantages – can be produced in large quantities and only one raw material needed. The process produces other useful products
- ...disadvantages – process requires huge amounts of electrical energy

3 a Salt raises blood pressure, which increases the risk of heart attacks and strokes.

b i Canada

ii People in different countries might have different tastes; there may be stricter regulations in some countries than others.

iii 6.4 g

iv This mass of salt is greater than the maximum daily UK recommended amount of salt for a 16-year-old. Pedro should aim to cut down his salt intake, and not eat two burgers in one day.

4 a Copper chloride, carbon dioxide, water

b Sodium nitrate and water

c Potassium sulfate and water

C4 Workout

1 a Column B: corrosive, toxic, oxidising

b Column C: wear gloves; use in a well-ventilated fume cupboard or wear mask over nose and mouth; keep away from flammable chemicals.

2 Life is fun. So is revision.

3 Picture of a snail.

4 She is beautiful, I think.

5 Red: group 2; blue: period 3; pencil: elements to right of stepped line (elements to the right of a stepped line between aluminium and silicon, germanium and arsenic, and so on); red circle: three from group 1, or hydrogen; blue circle: three from group 7

6

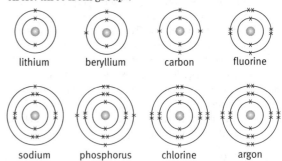

lithium beryllium carbon fluorine

sodium phosphorus chlorine argon

C4 Quickfire

1 Nucleus, neutrons, shells, electrons, electrons, 7, 7

2 a 7

b 7

c B

d 7

e 1

f 7

3 a Sodium + water $\longrightarrow$ sodium hydroxide + hydrogen

b Potassium + chlorine $\longrightarrow$ potassium chloride

c Hydrogen + iodine $\longrightarrow$ hydrogen iodide

d Lithium + water $\longrightarrow$ lithium hydroxide + hydrogen

e Sodium + chlorine $\longrightarrow$ sodium chloride

f Lithium + bromine $\longrightarrow$ lithium bromide

4 a 11 protons, 12 neutrons, 11 electrons

b 15 protons, 16 neutrons, 15 electrons

c 13 protons, 14 neutrons, 13 electrons

d 23 protons, 28 neutrons, 23 electrons

e 39 protons, 50 neutrons, 39 electrons

5

Name	Formula
water	H_2O
hydrogen gas	H_2
potassium chloride	KCl
sodium hydroxide	NaOH
iodine	I_2
chlorine gas	Cl_2
potassium bromide	KBr

6 a Reacts:

chlorine + sodium bromide $\longrightarrow$ sodium chloride + bromine

b No reaction

c Reacts:

bromine + sodium iodide $\longrightarrow$ sodium bromide + iodine

d No reaction

e Reacts:

chlorine + potassium iodide $\longrightarrow$ potassium chloride + iodine

7 a 2

b 3

c 6

d 2

e 1 atom of iron and 3 atoms of chlorine

8 a $2K(s) + 2H_2O(l) \longrightarrow 2KOH(aq) + H_2(g)$

b $2Na(s) + Cl_2(g) \longrightarrow 2NaCl(s)$

c $2Li(s) + 2H_2O(l) \longrightarrow 2LiOH(aq) + H_2(g)$

d $2K(s) + Cl_2(g) \longrightarrow 2KCl(s)$

e $2Fe(s) + 3Cl_2(g) \longrightarrow 2FeCl_3(s)$

9 a NaBr

b KCl

c MgS

d K_2O

10 a Sr^{2+}

b Be^{2+}

C4 GCSE-style questions

1 a Strontium is in group 2 of the periodic table, like calcium. So it is likely to react in a similar way to calcium.

b i 2.8.8.2

   ii 2

c i Any number between 715 and 849 is acceptable.

   ii Data from the column *reaction with water* shows that, as you go down the group, the reactions get more vigorous.

d Strontium + water $\longrightarrow$ strontium hydroxide + hydrogen

e i 1

   ii 2

   iii 4

2 a The melting points decrease as the group is descended.

b The prediction is incorrect. This decreases confidence in the explanation.

c 5/6 marks: answer clearly identifies the extent to which each set of data supports, or does not support, Ben's explanation **and** includes a reasoned assessment of whether or not the data, overall, do or do not support the explanation.

All information in the answer is relevant, clear, organised, and presented in a structured and coherent format. Specialist terms are used appropriately. There are few, if any, errors in grammar, punctuation, and spelling.

3/4 marks: answer identifies the extent to which one or both sets of data supports, or do not support, Ben's explanation **and** includes a brief assessment of the overall value of the data in supporting the explanation.

Most of the information is relevant and presented in a structured and coherent format. Specialist terms are usually used correctly. There are occasional errors in grammar, punctuation, and spelling.

1/2 marks: answer explains how one or other of the data sets supports, or does not support, the explanation **and** the explanation lacks detail/clarity. There may be limited use of specialist terms. Errors of grammar, punctuation, and spelling prevent communication of the science. Answer includes 1 or 2 points of those listed below.

0 marks: insufficient or irrelevant science. Answer not worthy of credit.

Relevant points include:

- Most of the data for the group 1 metal chlorides shows that as ion size increases, melting point decreases.
- The data for lithium does not fit this pattern.
- Overall, for the group 1 metal chlorides, the data supports the explanation.
- The data for the group 2 metal chlorides shows that as ion size increases, the melting point of the chloride increases.
- The group 2 metal chloride data does not support Ben's explanation at all.
- Overall, since one set of data supports the explanation and one set of data does not support the explanation, the data cannot be said to support Ben's explanation.

3 a i

Sodium	11	11	10	Na^+
Fluorine/fluoride	9	9	10	F^-

   ii

b The sodium and fluoride ions separate from each other and are free to move independently in the water.

c The sodium and fluoride ions are charged particles that are free to move independently.

4 a i 19

   ii 19

b Potassium

5 a NaBr

b $Cl_2(aq) + 2NaBr(aq) \longrightarrow 2NaCl(aq) + Br_2(l)$

C5 Workout

1 Box 1: A, oxygen ($O_2(g)$), argon (Ar(g)), carbon dioxide ($CO_2(g)$), nitrogen (N_2)(g)

Box 2: B, sodium chloride, potassium bromide, magnesium chloride, water

Box 3: C, silicon dioxide, aluminium oxide

2 Ab1; Ad1; Ai10; Ba2; Bh9; Bi10; Cc3; Ce6; Cf8; Ch4; Da5; Dg7

C5 Quickfire

1 a Oxygen

b Nitrogen

c Oxygen

d Oxygen

e Aluminium

2 CO_2: molecule D

H_2O: molecule A

O_2: molecule B

Ar: atom C

3 Carbon, covalent, four, covalent, hard, three, covalent, slippery, lubricant, can

4 a Zinc oxide + carbon ⟶ zinc + carbon dioxide
b Copper oxide + carbon ⟶ copper + carbon dioxide

5 a Iron oxide is reduced (red), carbon is oxidised (blue).
b Tin oxide is reduced (red), carbon is oxidised (blue).

6 A, C, B, D, E

7 Nitrogen: simple covalent: –196
Silicon dioxide: giant covalent: 2230
Sodium chloride: giant ionic: 1413

8 a $2ZnO(s) + C(s) \longrightarrow 2Zn(l) + CO_2(g)$
b $2Fe_2O_3(s) + 3C(s) \longrightarrow 4Fe(l) + 3CO_2(g)$
c $2CuO(s) + C(s) \longrightarrow 2Cu(l) + CO_2(g)$

9 a $Pb^{2+}(aq) + 2I^-(aq) \longrightarrow PbI_2(s)$
b $Cu^{2+}(aq) + 2OH^-(aq) \longrightarrow Cu(OH)_2(s)$

10 a 130 tonnes
b 27 kg

C5 GCSE-style questions

1 a Zn^{2+}
b Salt B
c Calcium bromide. Test with sodium hydroxide solution shows that salt A contains Ca^{2+} ions. Test with sodium hydroxide solution shows that the salt includes Br– ions.
d If hydrochloric acid were added, chloride ions would be present in the solution, so the solution would test positive for chloride ions, whether or not they were present in the original salt.
e $Ag^+(aq) + Cl^-(aq) \longrightarrow AgCl(s)$

2 5/6 marks: answer clearly explains why the properties of diamond make it suitable for drill tips **and** the explanation is logical and coherent.
All information in the answer is relevant, clear, organised, and presented in a structured and coherent format. Specialist terms are used appropriately. There are few, if any, errors in grammar, punctuation, and spelling.
3/4 marks: answer explains why the properties of diamond make it suitable for drill tips **but** the answer lacks detail **or** the explanation lacks logic and coherence.
Most of the information is relevant and presented in a structured and coherent format. Specialist terms are usually used correctly. There are occasional errors in grammar, punctuation, and spelling.
1/2 marks: answer briefly explains why the properties of diamond make it suitable for drill tips **and** the answer lacks detail **and** the explanation lacks logic and coherence.
There may be limited use of specialist terms. Errors of grammar, punctuation, and spelling prevent communication of the science. Answer includes 1 or 2 points of those listed below.
0 marks: insufficient or irrelevant science. Answer not worthy of credit.
Relevant points include:
- Diamond is very hard.
- So its surface cannot be worn down by other materials, which makes it a suitable material for drill tips.
- Diamond has a giant covalent structure.
- The covalent bonds between each carbon atom and its four neighbours are very strong.
- The strength of the covalent bonds means that they are difficult (require large amounts of energy) to break.
- Scratching diamond would involve breaking some covalent bonds at its surface.
- The large amounts of energy needed to break these bonds mean that diamond is very hard.

3 a i It has a lower boiling point than the other materials on the bar chart, showing that the forces that must be overcome in order to make the liquid boil are relatively weak.
ii Iodine
b i X or Z because they do not conduct electricity when solid.
ii Does the substance conduct electricity when liquid or in solution? If either X or Z does, then it is an ionic compound.
c i W
ii It conducts electricity when solid.

4 a i Hydrosphere
ii When solid, it does not conduct electricity – the charged particles cannot move; it has a high melting point – there are strong attractive forces between the ions; when solid it forms crystals – the ions are arranged in a regular pattern; when liquid, it conducts electricity – the charged particles can move independently.
iii $MgBr_2$
b 76.9 m^3
c i An electric current decomposes the electrolyte. An electric current passes through liquid magnesium chloride.
ii Magnesium is too reactive to be reduced by carbon – the ionic bonds in magnesium oxide are very strong.

5 a 176
b 14 g

6 a Giant, strong.
b i C
ii E

7 a i

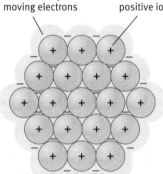

'sea' of freely moving electrons   lattice of positive ions

ii Aluminium is a good conductor of electricity because it has charged particles (electrons) that are free to move.
It is malleable because its layers of ions can slide over each other.
b i 529 t
ii 5/6 marks: all information in the answer is relevant, clear, organised, and presented in a structured and coherent format. Specialist terms are used appropriately. Few, if any, errors in grammar, punctuation, and spelling. Answer includes 5 or 6 points from those below.
3/4 marks: most of the information is relevant and presented in a structured and coherent format. Specialist terms are usually used correctly. There are occasional errors in grammar, punctuation, and spelling. Answer includes 3 or 4 points from those below.
1/2 marks: answer may be simplistic. There may be limited use of specialist terms. Errors of grammar, punctuation, and spelling prevent communication of the science. Answer includes 1 or 2 points from those below.

Points to include:
- Molten aluminium oxide is poured into an electrolysis cell.
- The cell has graphite electrodes.
- When a current flows through the liquid, positive aluminium ions move towards the negative electrode.
- Here they gain electrons to form neutral atoms.
- These atoms make up aluminium liquid, which is collected through a tap in the electrolysis cell.
- Negative oxide ions move towards the positive electrode.
- Here they lose electrons to become neutral atoms.
- Oxygen atoms join together in pairs to become oxygen molecules.

C6 Workout

1 Food additives e.g. saccharin; pharmaceuticals e.g. paracetamol; fertilisers e.g. ammonium nitrate; plastics e.g. polythene

2 **a** Petrochemicals and polymers
   b 38%
   c 4%

3 From left to right: B, C, A

4 1 acid or alkali; 2 titration flask; 3 solid sample; 4 burette; 5 pure water

5 1 desiccator, 2 crystallisation, 3 end point, 4 catalyst, 5 hydroxide, 6 slower, 7 sulfuric, 8 nitrate, 9 salt

C6 Quickfire

1

Gas	Liquid	Solid
hydrogen chloride	ethanoic sulfuric nitric	tartaric citric

2 **a** Acid **b** Alkali **c** Acid
   d Alkali **e** Both **f** Both
   g Both

3

Change	The reaction gets ...		
	faster	slower	can't tell
a Use bigger lumps of calcium carbonate.		√	
b Use more concentrated acid.	√		
c Heat the reaction mixture.	√		
d Use bigger lumps of calcium carbonate and heat the mixture.			√
e Add a catalyst.	√		

4 **a** Relative atomic mass is the mass of an atom of an element relative to the masses of other atoms.
   b An exothermic reaction is one which gives out energy / transfers energy to the surroundings.

5 96%

6

Name of chemical	Formula	Relative formula mass
nitrogen gas	N_2	28
nitric acid	HNO_3	63
magnesium sulfate	$MgSO_4$	120
potassium chloride	KCl	74.5
calcium chloride	$CaCl_2$	111
sodium carbonate	Na_2CO_3	106
calcium carbonate	$CaCO_3$	100

7

Formula of product	Actual yield	Theoretical yield	Percentage yield
SrO	98 kg	104 kg	94
Al_2O_3	222 g	224 g	99
SF_6	68 t	73 t	93

8 **a** +2
   b +3

9

Name of salt	Formula of acid used to make the salt	Formula of hydroxide used to make the salt	Formula of salt
potassium chloride	HCl	KOH	KCl
sodium sulfate	H_2SO_4	NaOH	Na_2SO_4
calcium nitrate	HNO_3	$Ca(OH)_2$	$Ca(NO_3)_2$
lithium chloride	HCl	LiOH	LiCl

1 **a** $NaOH + HCl \longrightarrow NaCl + H_2O$
      40 g   36.5 g   58.5 g   18 g
   b $2KOH + H_2SO_4 \longrightarrow K_2SO_4 + 2H_2O$
      112 g   98 g   174 g   36 g
   c $2Mg + O_2 \longrightarrow 2MgO$
      48 g   32 g   80 g
   d $4Li + O_2 \longrightarrow 2Li_2O$
      28 g   32 g   60 g
   e $AgNO_3 + NaCl \longrightarrow AgCl + NaNO_3$
      170 g   58.5 g   143.5g   85 g
   f $Pb(NO_3)_2 + 2KCl \longrightarrow PbCl_2 + 2KNO_3$
      331 g   149 g   278 g   202 g
   g $Fe_2O_3 + 3C \longrightarrow 3CO + 2Fe$
      160 g   36 g   84 g   112 g
   h $CaCO_3 \longrightarrow CaO + CO_2$
      100 g   56 g   44 g

C6 GCSE-style questions

1 **a** **i** $2HCl(aq) + CaCO_3(s) \rightarrow CaCl_2(aq) + CO_2(g) + H_2O(l)$
      ii 100 g of calcium carbonate produces 44 g of carbon dioxide, so 1.60 g of carbon dioxide is produced from 3.64 g of calcium carbonate.
   b Rate = 1.10 ÷ 1 = 1.10 g/min
   c **i** Experiment Y or experiment Z because both proceeded more slowly than the original experiment.
      ii Two from: increased temperature; increased acid concentration; used smaller pieces of calcium carbonate/calcium carbonate powder
   d Two from: the reaction is exothermic; the energy stored in the reactants is greater than the energy stored in the products; energy is released/given out in the reaction.

2 **a** Jude's data are more likely to give a value that is closest to the true value because he has collected a set of repeated values of the volume from which he can calculate a mean.
   b **i** Run 2, 11.90; run 3, 12.00; run 4, 12.10; run 5, 15.00
      ii Outlier is the value for run 5 (15.00 cm³).
      iii He had good reason to suspect that the outlier was inaccurate.
   c Mean = (11.9 + 12.00 + 12.10) ÷ 3 = 12.00 cm³
   d The results do not support the claim on the blackcurrant drink carton. The results show that the blackcurrant drink contains less vitamin C than orange juice, not four times as much.

3 **a** A F D B C E
   b **i** Water
       ii $H^+ + OH^- \longrightarrow H_2O$
       iii Citric acid
4 5/6 marks: answer clearly describes the steps for making copper sulfate crystals **and** includes a clearly explained suggestion of why Grace's yield is less than Nzila's for each step.
   All information in the answer is relevant, clear, organised, and presented in a structured and coherent format. Specialist terms are used appropriately. There are few, if any, errors in grammar, punctuation, and spelling.
   3/4 marks: answer describes the main steps for making copper sulfate crystals **and** for some of the steps gives clear explanations to explain why Grace's yield is less than Nzila's **or** for all of the steps gives brief reasons to explain why the yields might be different.
   Most of the information is relevant and presented in a structured and coherent format. Specialist terms are usually used correctly. There are occasional errors in grammar, punctuation, and spelling.
   1/2 marks: answer describes some of the steps for making copper sulfate crystals **but** does not suggest why the yields of the two experimenters might be different **or** the answer describes some of the steps for making copper sulfate crystals **and** suggests why the yields of the two experimenters might be different **and** the explanation lacks detail/clarity.
   There may be limited use of specialist terms. Errors of grammar, punctuation, and spelling prevent communication of the science. Answer includes 1 or 2 points of those listed below.
   0 marks: insufficient or irrelevant science. Answer not worthy of credit.
   Relevant points include:
   • Add copper oxide powder to sulfuric acid solution, with stirring, until no more will dissolved.
   • During this step, Grace might not have added enough copper oxide powder. This would have decreased her yield compared to Nzila.
   • Filter the mixture. Keep the solution.
   • During this step, Grace might not have waited until all the solution had passed through the filter paper. This would have decreased her yield compared to Nzila.
   • Heat the copper sulfate solution in an evaporating dish until its volume decreases to about half its original volume.
   • During this step, Grace might have allowed the solution to spit. This would have decreased her yield compared to Nzila.
   • Leave the concentrated solution in an evaporating dish to crystallise.
   • During this step, Grace might have spilt some of the solution, which would have decreased her yield compared to Nzila.
5 **a** 201 kg
   b 320 kg
   c 490 kg

C7.1 Workout

1 We reduce waste by developing processes with higher atom economies, finding uses for by-products, increasing recycling at every stage of the life cycle of a product.
   We make processes energy efficient by insulating pipes and reaction vessels and using energy from exothermic processes to heat reactants or generate electricity. We reduce pollution from wastes by removing or destroying

harmful chemicals before sending waste to air, water, or landfill sites, by neutralising hazardous acids and alkalis, or by precipitating toxic metal ions.
2 Total number of atoms in reactants = 9. Total relative atomic mass of these atoms = 173. Total number of atoms in calcium chloride = 3. Total relative atomic mass of these atoms = 111. Atom economy = 64%
3 –

C7.1 Quickfire

1 Ammonia, large, food flavourings, small, more, more
2 Bulk chemicals (red): ammonia, phosphoric acid, sodium hydroxide; fine chemicals (blue): food additives, fragrances, medicinal drugs.
3 Can the feedstocks be replaced as quickly as they are being used – if they can, the process will be more sustainable, since the feedstocks will not run out. The atom economy of the process – the higher the atom economy, the greater the proportion of reactant atoms that end up in the desired product. Waste products – the wastes, and what happens to them. Useful by-products – if the process produces useful by-products, the process is more sustainable. The environmental impacts – minimising damage to the environment makes a process more sustainable. Health and safety risks – if these are minimised, the process is more sustainable. Social benefits – greater social benefits increase the sustainability of a process. Economic benefits – greater economic benefits increase the sustainability of a process.
4 True statements: b, c.
   Corrected versions of false statements:
   a The activation energy of a reaction is the energy needed to break bonds to start a reaction.
   d The activation energy of a catalysed reaction is lower than the activation energy of the same reaction without a catalyst.
   e Many enzyme catalysts are denatured above temperatures of about 37 °C.
   f Enzyme catalysts only work at certain pH values.
5 **a** 100
   b 98
   c 158
   d 170
   e 159.5
   f 64
   g 261
   h 148
6 D, B, C, A
7 **a** $CuCO_3 \longrightarrow CuO + CO_2$
       123.5 g   79.5 g   44 g
   b $NaOH + HCl \longrightarrow NaCl + H_2O$
       40 g   36.5 g   58.5 g   18 g
   c $2Mg + O_2 \longrightarrow 2MgO$
       48 g   32 g   80 g
   d $2NaOH + H_2SO_4 \longrightarrow Na_2SO_4 + 2H_2O$
       80 g   98 g   142 g   36 g
   e $2HCl + CaCO_3 \longrightarrow CaCl_2 + CO_2 + H_2O$
       73 g   100 g   111 g   44 g   18 g
   f $2HCl + Mg \longrightarrow MgCl_2 + H_2$
       73 g   24 g   95 g   2 g
   g $Pb(NO_3)_2 + 2KI \longrightarrow PbI_2 + 2KNO_3$
       331 g   332 g   461 g   202 g
   h $2KOH + H_2SO_4 \longrightarrow K_2SO_4 + 2H_2O$
       112 g   98 g   174 g   36 g
8 8 g of copper oxide and 9.8 g of sulfuric acid
9 **a** 111 g
   b 44 g
10 1.7 g of silver nitrate and 1.5 g of sodium iodide

C7.1 GCSE-style questions

1 a i Yes – oranges are produced by the trees each year, and old orange trees can be replaced by planting new ones.

   ii The orange waste will no longer be used for animal feed.

   iii The maize for fuel is grown on land that might otherwise be used to grow food, but orange waste is the by-product of a crop (oranges) that is already being grown and a process (producing juice) that happens anyway. No new land is being used to produce the fuel.

  b New jobs will be created.

  c i Greenhouse gas and carbon monoxide emissions will be reduced.

   ii Emissions caused by transporting raw materials will be minimised.

   iii People might buy more of the cheaper fuel, so they will produce more greenhouse gases than they would have done with more expensive petrol.

  d Overall, the energy requirements of the process will be less.

  e i Carbon dioxide is a greenhouse gas that causes global warming/climate change.

   ii $C_6H_{12}O_6 \longrightarrow 2C_2H_5OH + 2CO_2$

   iii 37 °C

   iv The maximum concentration of ethanol solution that can be produced by fermentation contains around 14 % ethanol. The fuel must contain a greater concentration of ethanol than this.

2 5/6 marks: answer compares many advantages **and** disadvantages of each method **and** comes to a reasoned conclusion, based on evidence, about which method is more sustainable. All information in the answer is relevant, clear, organised, and presented in a structured and coherent format. Specialist terms are used appropriately. Few, if any, errors in grammar, punctuation, and spelling.

3/4 marks: answer compares some advantages and disadvantages of both methods **or** some advantages of both methods **or** some disadvantages of both methods **and** states which method is more sustainable **but** the reasons for the decision lack detail or do not refer to advantages and disadvantages identified. Most of the information is relevant and presented in a structured and coherent format. Specialist terms are usually used correctly. There are occasional errors in grammar, punctuation, and spelling.

1/2 marks: answer points out one or two advantages or disadvantages of one or both methods **but** does not state which method they believe to be more sustainable. There may be limited use of specialist terms. Errors of grammar, punctuation, and spelling prevent communication of the science.

0 marks: insufficient or irrelevant science. Answer not worthy of credit.

Relevant points include:

- Advantage of obtaining aluminium by recycling – less energy is required (15 MJ per kg of aluminium compared to 260 MJ per kg to obtain from its ore).
- Advantage of obtaining aluminium by recycling – no new ore is required, meaning the ore remains for the use of people in future.
- Disadvantage of obtaining aluminium by recycling – people who use aluminium have to make the effort to separate aluminium waste from their other rubbish.
- Disadvantage of obtaining aluminium by recycling – the lorries that collect the cans produce pollution near people's homes.

- Disadvantage of obtaining aluminium from its ore – large amounts of 'red mud' pollution produced.
- Health and safety risks – if these are minimised, the process is more sustainable.
- Social and economic benefits - greater social and economic benefits increase the sustainability of a process.

C7.2 Workout

1 Beer: ethanol; shampoo: an ester; pickled onions: ethanoic acid; butane camping gas: an alkane; car windscreen wash: made from methanol

2 Animal, energy, saturated, no more, esters, glycerol, fatty

3 Ab2; Af2; Ah2; Ah5; Ba3; Be1; Bh5; Ca3; Cf8; Dc7; De1; Dg4

C7.2 Quickfire

1 Ethanol – as a fuel; methanol – to make glue; ethanoic acid – to make vinegar; ethane – to make ethanol; pentyl pentanoate – as a food flavouring

2 Sweaty socks, rancid butter

3 Ethanoic acid – carboxylic acids; ethanol – alcohols; ethane – alkanes; ethyl ethanoate – esters

4 True statements: c, e.

Corrected versions of false statements:

  a Ethane burns in plenty of air to make carbon dioxide and water.

  b Alkanes do not react with acids because they contain only C–H and C–C bonds, which are difficult to break and therefore unreactive.

  d The limit to the concentration of ethanol solution that can be made by fermentation is about 14%.

  f *E. coli* converts waste biomass to ethanol.

  g A dilute solution of a weak acid has a higher pH than a solution of a strong acid of the same concentration.

  h Carboxylic acids are weaker acids than hydrochloric acid.

5

Name	Molecular formula	Structural formula
methane	CH_4	
butane	C_4H_{10}	
propane	C_3H_8	
methanol	CH_3OH	
methanoic acid	$HCOOH$	
ethanoic acid	CH_3COOH	

6 a Propane + oxygen ⟶ carbon dioxide + water

  b Ethanol + oxygen ⟶ carbon dioxide + water

  c Ethanoic acid + magnesium ⟶ magnesium ethanoate + hydrogen

  d Methanoic acid + magnesium oxide/hydroxide ⟶ magnesium methanoate + water

  e Ethanoic acid + calcium carbonate ⟶ calcium ethanoate + carbon dioxide + water

  f Ethanol + sodium ⟶ sodium ethoxide + hydrogen

7 a Ethanoic acid and ethanol

  b Propanoic acid and methanol

  c Ethanol and propanoic acid

  d Methanol and methanoic acid

8 Advantages: feedstocks are renewable; plants take in carbon dioxide from the atmosphere as they grow. Disadvantages: sugar cane is grown on land that could be used for food; carbon dioxide is emitted by tractors and in the manufacture of fertilisers used by sugar cane farmers.

9 Ethanol reacts with sodium to make sodium ethoxide and hydrogen. Ethane does not react with sodium, because it has only C–C and C–H bonds, which are difficult to break, and so unreactive. Water and ethanol react in a similar way with sodium. This is because both molecules have an O–H bond which breaks when they react with sodium.

10 A, E, C, D, B, F, G

11 a $CH_4 + 2O_2 \longrightarrow CO_2 + 2H_2O$

  b $2CH_3CH_2OH + 2Na \longrightarrow 2CH_3CH_2ONa + H_2$

  c $2CH_3OH + 3O_2 \longrightarrow 2CO_2 + 4H_2O$

  d $2C_2H_6 + 7O_2 \longrightarrow 4CO_2 + 6H_2O$

C7.2 GCSE-style questions

1 a To store energy

  b i A. It is unsaturated/contains double bonds between carbon atoms (C=C).

    ii The oil is suitable for frying potatoes to make crisps and, because oleic acid is unsaturated, it is less likely to increase a person's risk of getting heart disease.

2 a i C_2H_5OH

    ii Carboxylic acids

    iii To catalyse or speed up the reaction

  b Any two from: synthesis is cheaper; extraction from fruit is more complex; fruit takes a long time to grow; food crops need land to grow on.

3 5/6 marks: answer clearly describes the steps for making pure, dry ethyl ethanoate **and** includes a clearly explained suggestion of why Jamie's yield is less than Ester's for each step. All information in the answer is relevant, clear, organised, and presented in a structured and coherent format. Specialist terms are used appropriately. Few, if any, errors in grammar, punctuation, and spelling.

3/4 marks: answer describes the main steps for making ethyl ethanoate **and** for some of the steps gives clear reasons for Jamie's yield being less than Ester's **or** for all of the steps gives brief reasons to explain why the yields might be different. Most of the information is relevant and presented in a structured and coherent format. Specialist terms are usually used correctly. There are occasional errors in grammar, punctuation, and spelling.

1/2 marks: answer describes some of the steps for making ethyl ethanoate **but** does not suggest why the yields of the two experimenters might be different **or** the answer describes some of the steps for making ethyl ethanoate **and** suggests why the yields of the two experimenters might be different **and** the explanation lacks detail/clarity. There may be limited use of specialist terms. Errors of grammar, punctuation, and spelling prevent communication of the science.

0 marks: insufficient or irrelevant science. Answer not worthy of credit.

Relevant points include:

• Heat the mixture of reactants, and a concentrated sulfuric acid catalyst, under reflux.

• During this step, Jamie might not have heated for long enough, **or** some of the gaseous product made might not have condensed and returned to the reaction vessel.

• Distil the mixture obtained after reflux.

• Jamie might not have carried out the distillation process for a long enough time, **or** some of the gaseous ethyl ethanoate might not have condensed in the condenser, and so left the condenser as a gas and escaped into the air.

• Pour the ethyl ethanoate from the distillation stage into a separating funnel. Add dilute sodium hydroxide solution. Shake. Discard the aqueous layer.

• Add anhydrous calcium chloride granules to the purified ethyl ethanoate and distil the mixture again.

C7.3 Workout

1 Left diagram: an exothermic reaction; energy lost to surroundings; total energy of products is less than the total energy of reactants; examples: respiration and burning. Right diagram: an endothermic reaction; energy gained from surroundings; the energy of the products is more than the energy of reactants; example: photosynthesis.

2 a A and B

  b A and B

  c C

  d B

  e C and D

C7.3 Quickfire

1 Exothermic, less, endothermic, more

2 Exothermic – combustion and respiration; endothermic – photosynthesis and the reaction of citric acid with sodium hydrogencarbonate

3 a Exo b Exo c Endo d Endo e Both f Both
  g Exo h Endo

4 True statements: b, c
Corrected versions of false statements:

  a When molecules collide, they react if they have enough energy.

  d Each reaction has its own activation energy.

5 a A b D c Exothermic – the energy of the products is less than that of the reactants.

6 a Bonds that break are H–H and Cl–Cl; bonds that are made are two H–Cl bonds.

  b Bonds that break are C–H and Cl–Cl; bonds that are made are C–Cl and H–Cl.

7 a False

  b True

  c False

  d True

8 a −185 kJ   b −114 kJ   c −487 kJ

C7.3 GCSE-style questions

1 a The energy needed to break bonds to start a reaction

  b Vertical arrow pointing upwards, starting at dotted line and finishing at the top of the curve

2 5/6 marks: answer includes a correct calculation of the energy change for reaction 1 **and** a correct calculation for the energy change to make the same amount of hydrogen via reaction 2 **and** compares the two values. All information in the answer is relevant, clear, organised, and presented in a structured and coherent format. Specialist

terms are used appropriately. Few, if any, errors in grammar, punctuation, and spelling.

3/4 marks: answer includes a correct calculation of the energy change for reaction 1 **or** a correct calculation for the energy change to make the same amount of hydrogen via reaction 2 **and** compares the two answers. Most of the information is relevant and presented in a structured and coherent format. Specialist terms are usually used correctly. There are occasional errors in grammar, punctuation, and spelling.

1/2 marks: answer includes an incorrect attempt to calculate the energy change for reaction 1 **or** an incorrect attempt to calculate the energy required to make the same amount of hydrogen via reaction 2 **and** compares the two values. There may be limited use of specialist terms. Errors of grammar, punctuation, and spelling prevent communication of the science.

0 marks: answer not worthy of credit.

Relevant points include:
- The energy required to make the same amount of hydrogen via reaction 2 is $(286 \times 3) = 858$ kJ.
- The energy required to break the bonds in the reactants in reaction 1 is $(4 \times 413) + (2 \times 463) = 2578$ kJ
- The energy released on making the bonds in the products in reaction 1 is $(1077) + (3 \times 434) = 2379$ kJ
- The overall energy change for reaction 1 is $(2578 - 2379) = 199$ kJ
- The overall energy required for reaction 1 is approximately four times smaller than the energy required for reaction 2.
- All other factors being equal, the industry will choose the method with the smaller energy cost.

3 a Bonds broken: H–H and Cl–Cl. This requires $434 + 243 = 677$ kJ.
   b $(431 \times 2) = 862$ kJ
   c i $(677 - 862) = -185$ kJ
     ii Its value is negative.
     iii Arrow going down from reactants line to horizontal line labelled 'products'
   d The bond energy values given in the table are average values, so they may not be exactly the same as those for the molecules given in this question.

7.4 Workout

1 a Forward reaction: $PCl_5 \longrightarrow PCl_3 + Cl_2$; backward reaction: $PCl_3 + Cl_2 \longrightarrow PCl_5$; formulae of chemicals at equilibrium: PCl_5, PCl_3, Cl_2
   b Forward reaction: $N_2 + 3H_2 \longrightarrow 2NH_3$; backward reaction: $2NH_3 \longrightarrow N_2 + 3H_2$; formulae of chemicals at equilibrium: N_2, H_2, NH_3

2 Purified nitrogen gas from the air; purified hydrogen gas made from methane and steam; unreacted nitrogen and hydrogen are recirculated to increase the yield of ammonia; iron catalyst used to increase the rate of reaction; 450 °C – at higher temperatures, the yield decreases; 200 atm – at higher pressures, the yield increases; reaction vessel – equilibrium is not reached here because the gases do not stay in the reactor long enough to reach equilibrium; condenser – here ammonia gas condenses to the liquid state.

3 Forward reaction favoured by lower temperatures and higher pressures; backward reaction favoured by higher temperatures and lower pressures.

4 A2 (fixation); B6 (compound); C7 (legumes); D1 (nitrates); E3 (catalysts); F4 (triple bond); G5 (energy); H8 (fertilisers)

C7.4 Quickfire

1 True statements: a, e, f
  Corrected versions of statements that are false:
   b At equilibrium, the amounts of products remain the same.
   c Equilibrium can be approached from the product or the reactant side.
   d An equilibrium mixture contains products and reactants.
2 $H_2(g) + I_2(g) \rightleftharpoons 2HI(g)$
3 $CaCO_3(s) \rightleftharpoons CaO(s) + CO_2(g)$
4 Air, peas/beans, peas/beans, enzymes, ammonia, fixation
5 Hydrogen – steam and methane; nitrogen – air
6 As the pressure increases, the yield of ammonia increases.
7 450 °C, 200 atmospheres, iron catalyst
8 a I
  b I
  c D
  d I
9 Obtain hydrogen from water only instead of from methane and steam – to leave stocks of non-renewable resources for future generations; use a more efficient catalyst – to reduce energy use during production; supply energy from hydroelectric plants instead of from burning fossil fuels – to reduce greenhouse gas emissions
10 Chemists are searching for new catalysts that work in a similar way to natural enzymes to reduce the energy costs of the process.
11 Nitrates run into lakes and rivers. They make algae grow quickly. The algae damage ecosystems.
12 a As temperature increases, the yield of ammonia decreases.
   b Low temperatures increase the yield of ammonia. But at low temperatures the reaction is slow. The temperature chosen, 450 °C, is a compromise between the need to maximise both yield and rate.
13 The reaction shown is reversible. At equilibrium, NO_2 molecules are joining together all the time to make N_2O_4. At the same time, N_2O_4 is decomposing to make NO_2. The rates of the two reactions are the same. So the amounts of the two substances remain constant. This is dynamic equilibrium.

C7.4 GCSE-style questions

1 a Citric acid + water $\longrightarrow$ citrate ions + hydrogen ions
   b Citric acid, water, citrate ions, hydrogen ions
   c True statements: in solution, only some citric acid is ionised; in the equilibrium mixture, citrate ions and hydrogen ions react to make citric acid and water; in the equilibrium mixture, citric acid and water are reacting to make citrate ions and hydrogen ions.
2 a i The reaction is reversible.
     ii The gases do not stay in the reaction vessel long enough to reach equilibrium.
   b i To increase the yield of ammonia
     ii To increase the rate of reaction
   c i As temperature increases, yield decreases. The change is less rapid at higher temperatures.
     ii The reaction is exothermic, meaning it gives out energy. Le Chatelier's principle states that, when conditions change, an equilibrium mixture responds so as to counteract the effect of the change. At lower temperatures, more of the heat energy given out by the reaction can be absorbed, thus tending to counteract the effect of the change.

Answers

d The equation shows four gas molecules on the left and two on the right. Increasing the pressure shifts the equilibrium to the right, since this decreases the pressure, in line with Le Chatelier's principle.

3 5/6 marks: answer includes correct reasons for the choice of temperature **and** pressure **and** refers to data from the graph **and** equation. All information in the answer is relevant, clear, organised, and presented in a structured and coherent format. Specialist terms are used appropriately. Few, if any, errors in grammar, punctuation, and spelling.
3/4 marks: answer includes a correct reason for the choice of temperature **or** pressure **and** refers to data from the graph **or** the equation. Most of the information is relevant and presented in a structured and coherent format. Specialist terms are usually used correctly. There are occasional errors in grammar, punctuation, and spelling.
1/2 marks: answer includes a correct reason for the choice of temperature **or** pressure **or** refers to data from the graph **or** the equation in giving a partial reason for the choice of temperature **or** pressure. There may be limited use of specialist terms. Errors of grammar, punctuation, and spelling prevent communication of the science.
0 marks: answer not worthy of credit.
Relevant points include:
- The graph shows that the yield of sulfur trioxide at 450 °C is high.
- The rate of reaction would increase with temperature, but for this reaction the yield decreases at temperatures higher than 450 °C.
- The equation shows that 3 molecules of gas react to make 2 molecules of gas. This means that high pressures favour the forward reaction.
- The yield would increase as pressure increases.
- But atmospheric pressure is chosen for the reaction, perhaps because equipment and operating costs increase at higher pressures.

C7.5 Workout

1 From top: lid; solvent front; yellow – this dye moves up the paper faster; blue – this dye moves up the paper slower; paper – stationary phase; water – mobile phase and an aqueous solvent; equilibrium lies towards the right; equilibrium lies towards the left.

2 **a** A   **b** D   **c** A   **d** A

3 TLC: A; GC: C, E, H; paper chromatography: D; TLC and paper: B, F; all: G

4 **a** Left, from top: sodium hydroxide solution of known concentration, flask, indicator; middle, from top: pipette, hydrochloric acid; right: burette

b **i** 25.0 cm^3 **ii** 24.9–25.1 cm^3

C7.5 Quickfire

1 Qualitative analysis – to find out which chemicals are in a sample; quantitative analysis – to find out the amounts of the chemicals in a sample; store samples safely – to prevent samples being contaminated or tampered with; follows a standard procedure to collect samples – to be able to compare samples to each other or to standard results

2 C, A, D, B, E

3 **a** A solid sample cannot move up the chromatography plate.

b Pencil marks are not soluble in the solvent and so do not move up the chromatography plate; ink might be soluble in the solvent and interfere with the chromatography.

c To prevent evaporated solvent escaping

d When the chromatogram is removed from the chromatography tank, the solvent will evaporate, so the position of the solvent front cannot be seen. The distance the solvent front travels must be known in order to calculate values for R_f.

e To locate the spots, if they are not visible

4 X 0.3, Y 0.9

5 **a** Accurately weigh out 1.0 g of sodium hydroxide.

b Dissolve the sodium hydroxide in a small volume of pure water.

c Transfer the solution to a 250 cm^3 graduated flask.

d Rinse all of the solution from the beaker using more pure water.

e Add more water up to the 250 cm^3 mark of the graduated flask.

f Place a stopper in the flask and shake it.

6 **a** Jess's results will give a better estimate because the data have a smaller range which means they are more consistent and so closer to the true value.

b The mean of each set of data lies in the range of the other set of data.

7 Copper sulfate 4 g/dm^3; sodium chloride 30 g/dm^3; magnesium sulfate 300 g/dm^3; zinc bromide 160 g/dm^3

8 **a** 40 g
b 1.5 g
c 3 g
d 50 g

9 **a** 4.7 g/dm^3
b 12.1 g/dm^3

C7.5 GCSE-style questions

1 **a** **i** To sterilise his skin so the needle did not cause an infection. The wipe was alcohol free because using a wipe that contains alcohol could increase the amount of alcohol in the sample.

ii To ensure that the sample was not muddled with that of anyone else

iii To check that the sample had not been tampered with

iv To make sure the sample did not 'go off'

b **i** 1.0 minute
ii Four (one not named)

c No, it does not support Ron's belief that the results of the first analysis are incorrect because the printouts from the two analyses are identical.

2 5/6 marks: the answer identifies **and** compares the compounds present in each fruit juice **and** compares their amounts. All information in the answer is relevant, clear, organised, and presented in a structured and coherent format. Specialist terms are used appropriately. Few, if any, errors in grammar, punctuation, and spelling.
3/4 marks: the answer identifies **or** compares the compounds present in each fruit juice **and** compares their amounts **or** the answer identifies and compares the compounds present in each fruit juice **but** does not compare their amounts. Most of the information is relevant and presented in a structured and coherent format. Specialist terms are usually used correctly. There are occasional errors in grammar, punctuation, and spelling.
1/2 marks: the answer identifies the compounds present in one type of juice but does not compare their amounts. There may be limited use of specialist terms. Errors of grammar, punctuation, and spelling prevent communication of the science.
0 marks: answer not worthy of credit.
Relevant points include:
- Five compounds were detected in cranberry juice, and four in blackcurrant juice.

- Cranberry juice contains compounds A, B, D, E, and G.
- In cranberry juice, there is a greater amount of compound D than of the other compounds.
- Blackcurrant juice contains compounds B, C, E, and F.
- In blackcurrant juice, there is a greater amount of compound C than of any other compound, and also a relatively large amount of compound F.

Ideas about science 1 Workout

1 1 data, 2 range, 3 outlier, 4 repeatable, 5 melting point, 6 inaccurate, 7 mean, 8 higher, 9 true 10 repeat, 11 lowest, 12 solid, 13 opinions, 14 no, 15 silver

2 **a** A   **b** B   **c** D   **d** A and C

Ideas about science 1 GCSE-style questions

1 **a** **i** To gather a set of data from which to calculate a mean value. The value obtained will then be as close as possible to the true value.
  ii It was difficult to know exactly when the end point was reached; any other sensible suggestions
 b 11.3
 c **i** The first reading of 12.0 cm^3
  ii The titration reading was a rough one, which Ben did to get a rough idea of the volume of acid required.
 d 11.3–11.5 cm^3
 e 11.4 cm^3
2 **a** **i** 5
  ii There is no specific reason to doubt its accuracy.
 b Range = 1.2 to 1.6 g; mean = 1.4 g
 c The mean for restaurant B is outside the range of the mean for restaurant A.
 d The data for restaurant B is reproducible.
3 5/6 marks: answer gives correct values of the mean and range for each car **and** points out that the mean for car A is within the range of car B, and vice versa **and** clearly states that this means there is no real difference between the carbon dioxide emissions for the two cars. All information in the answer is relevant, clear, organised and presented in a structured and coherent format. Specialist terms are used appropriately. Few, if any, errors in grammar, punctuation, and spelling.
3/4 marks: answer correctly gives the values of the means for both cars **or** the ranges for both cars **or** one of each **and** points out that the mean for car A is within the range of car B **or** the mean for car B is within the range of car A, **or** states that there is no real difference between the carbon dioxide emissions for the two cars. Most of the information is relevant and presented in a structured and coherent format. Specialist terms are usually used correctly. There are occasional errors in grammar, punctuation, and spelling.
1/2 marks: answer correctly gives the values of the range for one car **or** the mean for one car **and** states that there is no real difference between the carbon dioxide emissions for the two cars. There may be limited use of specialist terms. Errors of grammar, punctuation, and spelling prevent communication of the science. Answer includes 1 or 2 points of those listed below.
0 marks: insufficient or irrelevant science. Answer not worthy of credit.
Relevant points include:
- The mean for car A is 158 g/km.
- The mean for car B is 160 g/km.
- The range for car A is 153 – 163 g/km.
- The range for car B is 156 – 164 g/km.
- The mean for car A is within the range of car B, and the mean for car B is within the range of car A.
- This means that there is no real difference between the carbon dioxide emissions for the two cars.
4 **a** Wind direction varied; a nearby coal-fired power station was running on some days, but not on others; one student used the measuring equipment incorrectly (1 mark if all three correct answers are given).
 b The percentage of damaged sperm in samples taken from men at regular intervals throughout the six months.

Ideas about science 2 Workout

1 A, B, C, E
2 **a** The average mass of citrulline in water melon flesh is lowest in red watermelons and highest in yellow ones.
 b On average, men have bigger ears than women.
 c For men, ear size increases with age.
 d As the average light intensity increases, the number of foxgloves growing in a 1 m^3 plot decreases.
 e As the number of generator coils increases, the voltage increases.
 f As temperature increases, the time to collect 100 cm^3 of gas decreases.

Ideas about science 2 GCSE-style questions

1 **a** Factor: type of shampoo; outcome: percentage of breakage
 b **i** Person the hair is from, length of hair
  ii To make sure the test is fair
 c There could be some other factor that made the hair that was washed with the anti-break shampoo stronger.
2 **a** Factors: A, F; outcomes: B, C, E
 b 1 – C; 2 – D; 3 – A; 4 – B
3 **a** The smaller the concentration of ozone in the atmosphere, the greater your chance of getting a cataract; there is a correlation between the concentration of ozone in the upper atmosphere and the number of people with cataracts; wearing sunglasses that protect against ultraviolet radiation may reduce your chance of getting a cataract.
 b **i** As large as practically possible
  ii Time people spend outdoors; ethnic origin
4 **a** Graph 1 – reaction D; graph 2 – reaction B or reaction C; graph 3 – reaction A
 b **i** Percentage of sulfur trioxide in the equilibrium mixture
  ii Three factors that might affect the outcome: pressure, temperature, amounts of sulfur dioxide and oxygen in starting mixture. The chemist must control temperature and amounts of sulfur dioxide and oxygen in starting mixture, because these would affect the outcome if not controlled, making the investigation invalid.
 c As the percentage of catalyst increases, so the percentage of sulfur dioxide that is converted to sulfur trioxide increases.
For each percentage of catalyst, at first the percentage of sulfur trioxide rises rapidly as the percentage of oxygen increases. In all cases, the percentage of sulfur trioxide reaches a maximum when the percentage of oxygen is between 1% and 2%.
 d Answers might include statements such as:
  - Carrie's statement is incorrect: in fact, the higher the temperature, the smaller the percentage that is converted.

- Joe's statement is also incorrect: some factor other than temperature change could cause the change in percentage of sulfur dioxide that is converted.
- Leah's statement is partly correct, in that, for the temperatures shown on the graph the percentage converted is higher at lower temperatures. But the graph does not show the percentage converted below 300 °C.
- Ezekiel's statement is incorrect: the graph does not give any information about catalysts.

Ideas about science 3 Workout

1 Statements that are true: c, d, f, h
   Corrected versions of statements that are false:
   a Scientific explanations have to be thought up creatively from data.
   b An explanation may be incorrect, even if all the data agree with it.
   e An explanation may explain a range of phenomena that scientists didn't know were linked.
   g If an observation agrees with a prediction that is based on an explanation, it increases confidence that the explanation is correct.
   i If an observation does not agree with a prediction that is based on an explanation, then the explanation or the observation may be wrong. Confidence in the explanation is reduced.
   j If an observation does not agree with a prediction that is based on an explanation, then the explanation or the observation may be wrong. Confidence in the explanation is reduced.
2 1 A and C; 2 E; 3 B; 4 D
3 1 A, B, C, D; 2 E; 3 F; 4 G

Ideas about science 3 GCSE-style questions

1 a He was not there when life began so could not observe what happened (or any other sensible answer).
   b Two correct statements are: if the prediction is correct, we can be more confident that the explanation is correct; if the prediction is wrong, we can be less confident that the explanation is correct.
   c Yes. The experiments simulated the conditions described by Explanation 1, and the chromatogram shows that amino acids were formed, as stated in the explanation.
   d i A and C
      ii The scientists could do an experiment based on Explanation 2, simulating the conditions of a hydrothermal vent and analysing the products to find out whether amino acids are formed. If they are not, this might suggest that explanation 1 is better.

Ideas about science 4 Workout

1 1 E, 2 C, 3 B, 4 D, 5 G, 6 A, 7 F, 8 H

Ideas about science 4 GCSE-style questions

1 a The result made other scientists less likely to accept Masataka Ogawa's claim; the result made other scientists more likely to question Masataka Ogawa's claim.
   b i To gain more evidence so that other scientists have greater confidence in their claim.
      ii The results would make other scientists more likely to accept Ida and William's claim.

Ideas about science 5 Workout

1 1 harm; 2 chance; 3 advance; 4 ionising; 5 consequences; 6 assess; 7 controversial; 8 statistically; 9 benefit; 10 perceive; 11 acceptable

2 Sentences might include the following:
   - People's perceptions of the size of a particular risk may be different from the statistically estimated risk.
   - The perceived risk of an unfamiliar activity is often greater than the perceived risk of a more familiar activity.
   - Governments have to assess what level of risk is acceptable in a particular situation. This decision may be controversial.

Ideas about science 5 GCSE-style questions

1 a 400 g
   b i The amount of salt in the bread and spread is $(1.5 \times 0.75) + (0.1 \times 1.4) = 1.265$ g
         The amount of salt in the rice flakes and milk is $(1.2 \times 0.5) + (0.04 \times 2) = 0.68$ g
         So Oliver should have the rice flakes and milk.
      ii The oats and milk both have tiny amounts of salt in them.
   c Pressure from health organisations and the government have persuaded manufacturers to reduce the amounts of salt in their products.
   d The government may not want to upset the food industry; the government may believe it is up to people to make their own decisions about whether or not to eat salty food.

Ideas about science 6 Workout

1 a

A person or people who identify...	Name or names
...an impact on the environment.	Sharita
...an issue that science cannot solve.	Clarence
...issues that could be investigated scientifically.	Karyl, Sharita
... an ethical issue.	Clarence
...an unintended impact on the environment.	Sharita
...an issue linked to sustainability.	Paulina

   b i

	Benefits	Drawbacks
Extracting aluminium from bauxite ore	jobs for people in bauxite mines	Red mud waste damages environment; higher energy costs; more carbon dioxide; air pollution causes health problems.
Recycling aluminium	lower energy costs; less red mud pollution; less air pollution to cause breathing problems; recyclers make money; less carbon dioxide during production	Some people find recycling aluminium a nuisance; fuel required to transport aluminium for recycling.

   ii *Paragraph could include ideas such as those below, as well as an opinion as to which method of aluminium production is better.*
       The benefit of extracting aluminium from bauxite ore is that it provides jobs for local people. However, there are many drawbacks to extracting bauxite from its ore, including environmental costs such as the red mud pollution produced, and the production of carbon dioxide. Extracting bauxite from its ore produces air pollution which may cause health problems, and has higher energy costs than producing aluminium by recycling.

Recycling aluminium produces less pollution than extracting aluminium from its ore. The process also leads to smaller carbon dioxide emissions. Aluminium recycling provides jobs. As well as the benefits described, aluminium recycling does have some disadvantages. For example, some people find that recycling aluminium is a nuisance. Fuel is required for the lorries that transport scrap aluminium to recycling plants.

Ideas about science 6 GCSE-style questions

1 a Local people may get jobs; the mining company might make a profit.
  b Local roads might be damaged; tiny particles of graphite might pollute the air near the mine.
  c What mass of graphite is in 1 tonne of rock from the mine?; is the graphite of good enough quality to make batteries?; how does graphite dust affect health?
  d i About 800 years
    ii Recycling graphite means that more will be left in the Earth's crust for future generations.
  e i Graphite dust can damage health, and if there were no regulations limiting exposure levels, companies might expose people to dangerously high levels.
    ii Harm done to people who have been exposed to different levels of graphite dust in the past.

Index